IAEI

International Association of Electrical Inspectors

SOARES BOOK ON
GROUNDING

J. PHILIP SIMMONS
AUTHOR

SIXTH EDITION

Notice to the Reader

Publisher does not warrant or guarantee any of the products described herein or perform any independent analysis in connection with any of the product information contained herein. Publisher does not assume, and expressly disclaims, any obligation to obtain and include information referenced in this work.

The reader is expressly warned to consider carefully and adopt all safety precautions that might be indicated by the activities described herein and to avoid all potential hazards. By following the instructions contained herein, the reader willingly assumes all risks in connection with such instructions.

The publisher makes no representations or warranties of any kind, including but not limited to, the warranties of fitness for particular purpose or merchantability, nor are any such representations implied with respect to the material set forth herein, and the publisher takes no responsibility with respect to such material. The publisher shall not be liable for any special, consequential or exemplary damages resulting, in whole or in part, from the readers use of, or reliance upon, this material.

Published by:
International Association of Electrical Inspectors
901 Waterfall Way, Suite 602
Richardson, TX 75080-7702
(800) 786-4234

Copyright© 1996 by International Association of Electrical Inspectors.

Sixth edition, First printing.
Printed in the United States of America.

Library of Congress Cataloging in Publication Data

Simmons, J. Philip
     IAEI Soares book on Grounding, 6th edition
     Includes index.

Library of Congress Catalog Card Number: 96-77947

Author   J. P. Simmons
Managing Editor   M. V. Cadungog

Dedicated to the
Memory
of
Eustace C. Soares, P.E.

Preface

Eustace Soares' book, "Designing Electrical Distribution Systems for Safety" was originally published in 1966 and was based upon the 1965 edition of the National Electrical Code® (NEC®). Over the years, this book has become a classic.

Eustace was one of the most renowned experts in the history of the National Electrical Code® in the area of grounding electrical systems. A wonderful teacher and man of great vision, Eustace foresaw the need for better definitions to clear up the great mystery of grounding of electrical systems.

A great majority of the recommendations contained in the original edition of his book have been accepted as part of Article 250 of the National Electrical Code®. The grounding philosophies represented in the original edition are just as relevant today as they were then. To say that Eustace contributed more than any other man to solving some of the mysteries of grounding of electrical systems would not be an overstatement of fact.

Previous editions have been extensively revised both in format and in information. An effort has been made to bring the work into harmony with the 1996 edition of the National Electrical Code®. This author has attempted to retain the integrity of the technical information for which this work has been well known and at the same time add additional information which may be more recent on the subject of grounding.

It is our intent to revise this work to complement each new National Electrical Code® so this will be an on-going project. If the reader has suggestions for additional pertinent material or comments about how this work could be improved upon, they would be most welcome.

National Electrical Code® and NEC® are registered trademarks of the National Fire Protection Association, Inc., Quincy, MA 02269.

About the Author

J. Philip Simmons is the author of this sixth edition of this text and was responsible for updating the work to the 1996 National Electrical Code®. He also was responsible for updating the fourth and fifth edition of this text and for completely reorganizing and expanding the work.

Mr. Simmons presently serves as director of codes and education for the International Association of Electrical Inspectors. As such, he is responsible for developing and updating educational materials. He also serves as a presenter at seminars on electrical codes and safety. Phil served as executive director of IAEI for five years and was international president in 1987 after serving on the board of directors. Other service to the IAEI by Simmons has included Secretary-Treasurer and Chairman of the Puget Sound Chapter and Secretary-Treasurer of the Northwestern Section.

Simmons is contributing editor of the *IAEI News* and was author of the IAEI books on the Analysis of the 1993 and 1996 National Electrical Code®. He was coauthor of the IAEI book on electrical systems in One- and Two-Family Dwellings and has served as an instructor at many seminars throughout the country. Phil is author of the 1996 and 1993 editions of Ferm's Fast Finder Index book. He also is chairman of the Neon Installation Manual committee.

Formerly chief electrical inspector of the state of Washington, Department of Labor and Industries, Mr. Simmons also served the people of the state as a field electrical inspector, electrical plans examiner, health care facility inspector and electrical inspector supervisor.

Mr. Simmons worked in an industrial plant for several years where he learned the electrical trade. He went on to form his own electrical contracting firm which he successfully operated for several years. Simmons is licensed as a journeyman electrician and administrator in the state of Washington. He successfully passed the certification examinations for construction code inspectors and holds a certificate in the One- and Two-Family Electrical Inspector, General Electrical Inspector and Electrical Plan Review classifications.

A student of the NEC® for many years, Mr. Simmons developed and taught Code related courses for the state of Washington and the IAEI. He served as an instructor for the IAEI, teaching the analysis of the changes in the National Electrical Code® and grounding of electrical systems for safety.

Mr. Simmons presently serves on the NFPA, National Electrical Code®, Code Making Panel 5 which is responsible for Article 250, Grounding. He has served on CMP-1 and 17 and as chairman of Code Making Panel 19. He also is a member of the NEC® Technical Correlating Committee.

Simmons served for six years as a member of the NFPA Standards Council and is past chairman of the NFPA Electrical Section. Other service to NFPA include the Advisory Committee on Adoption and Use of the National Electrical Code® and Related Documents and the Bi-National Correlating Committee for the Harmonization of the Electrical Codes.

Mr. Simmons serves on the Underwriters Laboratories' Electrical Council and is a past UL Trustee. He also is a member of the International Brotherhood of Electrical Workers.

He also is a member of the electrical advisory committee, National Certification Program for Construction Code Inspectors for The Chauncey Group, formerly the Educational Testing Service.

Phil and his wife Dorothy have five children and ten grandchildren. His education includes a certificate in Public Administration and many developmental and continuing education programs.

Table of Contents

Chapter 1

General

Objectives

After studying this chapter, the reader will be able to understand:

- Fundamentals and purpose of grounding of electrical systems.
- Various definitions relative to grounding.
- Equipment from grounded and ungrounded systems.
- Effect of electric shock hazards on humans and animals.
- Purpose of grounding and bonding.
- Differences between a short circuit and ground fault.
- Basic design requirements for grounding of electrical systems.
- Circuit impedance and other characteristics.

Grounding of electrical systems.

Some features of electrical safety are so fundamental they have appeared in some form in every edition of the National Electrical Code®. These include suitable insulation for conductors and overcurrent protection for circuits.

Another long-time common electrical safety code requirement is grounding of electrical systems and equipment for safety. The grounding of equipment and conductor enclosures, as well as the grounding of one conductor of an electric power and light system, has been practiced in some quarters since the use of electricity began.

At first, there was no uniform standard for grounding. (See the Appendix of this text for information on the early history of grounding including debates that took place on how best to approach the subject.) However, it was not long before it became universal to ground one conductor on all 120-volt lighting circuits. Early code books firmly established the practice of grounding by making it mandatory to ground all such systems where the voltage to ground does not exceed 150 volts. These code books also recommended that alternating current systems be grounded where the voltage to ground did not exceed 300 volts while at the same time stating that higher voltage systems were permitted to be grounded. Later, the Code made it mandatory to ground any system having a nominal voltage to ground of not more than 300 volts if the grounded service conductor (usually a neutral) was not insulated.

The Earth as a Conductor

Grounded: "Connected to earth or to some conducting body that serves in place of the earth."*

The earth as a conductor is assumed to have a potential of zero.

Figure 1-1

Definition.

The term "grounded," is defined in the National Electrical Code® (NEC®) in Article 100 as "Connected to earth or to some conducting body that serves in place of the earth."* Conducting bodies that serve in place of the earth include conduit, enclosures and equipment grounding conductors that are, in fact, an extension of the earth by being electrically connected to the earth by reliable electrical and mechanical means.

The earth as a whole is properly classed as a conductor. For convenience, its electric potential is assumed to be zero. Based upon the composition of

Grounding Fundamentals

Equipment connected to earth - potential equalized.

Earth

Apply potential to equipment from fault - current flow to source. (From grounded system)

Earth

Apply potential to equipment from fault - potential difference. (From grounded system)

(Ungrounded equipment)

Earth

Figure 1-2

the earth, the resistance of the earth can vary widely from one area to another. The earth is composed of many different materials, some of which, especially when dry, are very poor conductors of electricity. Soil temperature and moisture content are other factors that have a great influence on soil resistance. As a result, the capability of the earth to carry electrical current also varies widely. See Chapter Six of this text for more information on composition of the earth.

Where a metal object is grounded by connecting it to the earth by means of a grounding electrode, grounding electrode conductor and/or equipment grounding conductor, it too is thereby forced to take the same zero potential as the earth. Any attempt to raise or lower the potential of the grounded object results in current passing over the grounding connection until the potential of the object and the potential of the earth are equalized. Hence, grounding is a means for ensuring that the grounded object cannot take on a potential differing enough from earth potential to be hazardous.

Equipment from grounded system.

Where the electrical system is grounded, it is critical to provide a low-impedance path of adequate capacity from all equipment supplied by the system back to the source of the system. This is required by Section 250-51 and is emphasized in several portions of this text.

Where electrical equipment that is supplied by a grounded system is left ungrounded, it becomes a silent and often lethal source of electrical shock where a ground-fault occurs.

Another hazard of equal significance is where one piece of electrical equipment is grounded and another piece of equipment within reach of the person is not grounded and becomes energized through a failure of the insulation system. The person making contact with the two (or more) pieces of equipment becomes the circuit for current to pass through as it tries to find its way back to its source. In some cases, the person will receive a mild shock. In other cases, the shock may be fatal.

Equipment from ungrounded system.

Section 250-5 permits some electrical systems to be operated ungrounded. In this case, the electrical enclosures for the service, feeders, circuits and other equipment are connected to ground but the system itself is not grounded.

From a practical standpoint, an ungrounded

system exists only in theory or at the distribution transformers hanging on poles before they are connected to the plant electrical system. Where the ungrounded system conductors are installed in metal raceways and enclosures and connect to motors, the ungrounded system becomes capacitively coupled to ground.

Ungrounded electrical equipment which becomes energized from an ungrounded system conductor becomes energized at the source voltage potential to ground. Any person or animal coming into contact with the energized equipment becomes a conductor for the ground-fault current to flow back to the source of the system. The voltage of the source and the resistance of the current path provided by the person or animal determine the amount of current that will flow in the circuit.

Effect of electricity on humans.

Like the bird on the electric wire, the human body is immune to electric shock as long as it is not a part of the electric circuit. The easiest way to avoid danger from electric shock is to keep ones body from becoming a part of the electric circuit. Due to the common use of electric tools, equipment and appliances, the risk of being exposed to electric shock is multiplied by the number of this equipment we are exposed to.

When a person becomes a path for electricity, the person will experience an electrical shock. The intensity and damage done to the person by the shock will be determined by the current level (the amount that flows through the person), how long the current flows (the duration) the person's size and the pathway the current takes through the body.

Human Completing Grounding Circuit

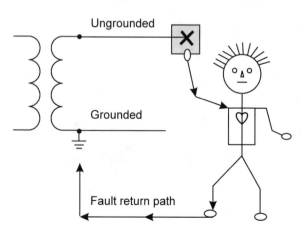

Figure 1-3

A person may become a pathway to ground in one of two ways; in a series or a parallel circuit. In a series circuit, the person is the only path through which the current will flow in its attempt to return to its source. An example of this is when a person comes into contact with a faulted electrical appliance and at the same time touches a grounded appliance like an electric range or a grounded kitchen sink.

Human in Parallel Ground-Fault Circuit

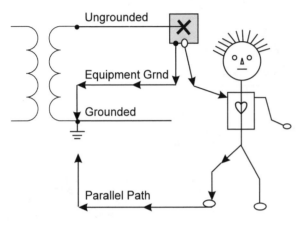

Figure 1-4

In a parallel circuit, the human and another path for current flow exist at the same time. In this case, usually not enough current is flowing through the equipment grounding conductor to cause the overcurrent protective device to clear the faulted equipment. This can be due to the equipment grounding conductor being sized too small or the equipment being energized through a resistance. Another factor is the time it takes for the overcurrent device to clear the fault from the circuit. During this time, the equipment subject to the ground-fault will have a potential above ground depending upon the voltage drop of the circuit.

Our primary emphasis is to provide a permanent, adequately sized and low-impedance path from all electrical equipment back to the source to facilitate the operation of the overcurrent device in a reasonable time. Part of the ground-fault current flows through the person who contacts the faulted equipment and the earth or grounded surfaces.

The protection of persons from electric shock is the reason ground-fault circuit-protective devices have become so popular over the recent years. See Chapter 14 of this text for additional information on the subject.

Effect of electricity on animals.

Some animals are especially sensitive to electricity. For example, past studies claimed that dairy cattle are so sensitive to electricity that a potential of as little as two volts between conductive portions of floors, walls, piping and stanchions caused behavior problems that resulted in loss of production. More recent studies claim the voltage difference to be about four volts. In other cases, severe health problems are attributed to the effects of electricity, which if not corrected and treated may lead to death. See Chapter 16 of this text for methods of preventing and correcting these problems in agricultural structures.

Taking the mystery out of grounding.

For many years the subject of "Grounding" has been considered one of the most controversial and misunderstood chapters in the National Electrical Code®. There is no real reason why the subject should be treated as a mystery and given so many different interpretations.

Probably the single most important requirement for clearing up the confusion about grounding is to clearly review and understand the definitions of the various elements of the grounding system found in the National Electrical Code®. In addition, these terms should be correctly used during discussion and instruction on the subject of grounding so all will have a common understanding. For example, using the term "ground wire" to mean an "equipment grounding conductor" does no more to help a person understand what specific conductor is referenced than the use of the term "vehicle" to specifically mean a "truck."

Carefully review the grounding terms that are defined at the beginning of each chapter in this text for clear understanding. In many illustrations throughout the text, portions of the grounding system are pointed out for assistance in clearer understanding.

Purpose of grounding and bonding.

Systems are solidly grounded to limit the voltage to ground during normal operation and to prevent excessive voltages due to lightning, line surges or unintentional contact with higher voltage lines and to stabilize the voltage to ground during normal operation.

Equipment grounding conductors are bonded to the system grounded conductor to provide a low-impedance path for fault current that will facili-

Purpose of Equipment Grounding Conductor

Figure 1-5

tate the operation of overcurrent devices under ground-fault conditions. (Section 250-1 FPN No. 1)

Solid system grounding facilitates the automatic clearing of ground faults by circuit-protective equipment. Other types of system grounding permit the functioning of relays to operate an alarm to indicate that the electrical system has developed a line-to-ground fault.

Purpose of Bonding Enclosure

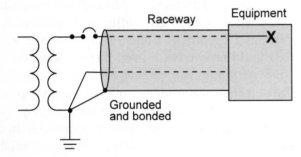

Figure 1-6

Conductive materials enclosing electrical conductors or equipment or that are a part of the equipment are grounded to limit the voltage to ground on these materials and bonded to facilitate the operation of overcurrent devices under ground-fault conditions. (Section 250-1 FPN No. 2) Several methods of grounding electrical systems are used depending on National Electrical Code® requirements and system design and function. These methods of grounding electrical systems are covered in detail in Chapter Three of this text.

Scope of grounding.

The "scope of grounding" is contained in Section 250-51 of the NEC® which is headed "Effective Grounding Path." This section requires that "The

path to ground from circuits, equipment, and metal enclosures for conductors shall:

1. be permanent and electrically continuous;
2. have capacity to conduct safely any fault current likely to be imposed on it; and
3. have sufficiently low impedance to limit the voltage to ground and to facilitate the operation of the circuit-protective devices."*

Scope of Grounding

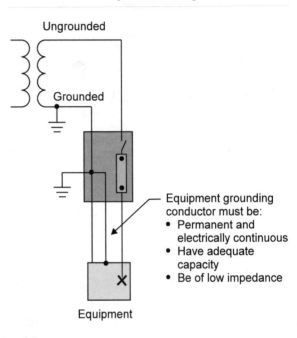

Figure 1-7

Each of these points is important and worthy of discussion. Permanent, reliable and electrically continuous grounding systems are vital to safety of the electrical system. Intermittent connections are like an unpredictable earthquake, waiting to wreak havoc on the unsuspecting. The grounding system, including all connections in the system, must be permanent and electrically continuous.

Providing adequate capacity for the fault current that will flow in the circuit is also of paramount importance. The minimum size system grounded conductor, bonding conductor and equipment grounding conductor are given in several locations in the NEC®, many of them in Article 250. Obviously, a grounding conductor that burns off due to excessive current flow while it is carrying out its intended purpose is of little value in the safety system.

Conductors that are permanently connected and of adequate capacity are of little value if they do not have low impedance. A high impedance (measured in ohms) path provides resistance to the system thus limiting the amount of current that can flow over the circuit. This allows a voltage above ground to be present on faulted equipment which may present a shock hazard. The Code stops short of specifying the maximum impedance acceptable for circuits. See Chapter Eleven of this text for additional information on the subject.

The goal of the ground-fault path is to provide a permanent and adequate path of low impedance so enough current will flow in the circuit to cause a circuit breaker to trip or a fuse to open. If the opening of the breaker or fuse on the line side of the faulted circuit does not take place in a rapid manner, thus taking the faulted equipment off the line, the grounding system will have failed to perform its function.

Electrical system design.

The ultimate goal of electrical system protection is to prevent ground faults from occurring by means of proper design, installation, operation and maintenance of electrical equipment and systems. However, the probability of failure or incorrect connections will exist in any electrical installation. Therefore, ground-fault protection must be provided to safely clear line-to-ground faults which may occur.

Section 110-7 requires that, "All wiring shall be so installed that, when completed, the system will be free from short circuits and from grounds other than as required or permitted by Article 250."* Though not widely practiced, it is wise to test the circuit to be certain it is free from unintended short circuits and/or ground faults before it is energized. This can be done with a battery, bell and leads (the origin of the term "ring-out"), an ohmmeter or megohm-meter (megger).

Electrical systems abused.

Electrical systems are intended to provide reliable service for many years. That is generally the case, provided the system is designed for the load to be carried by the system and is installed in a proper manner. Overloading the system is one of the major abuses it must endure. With the expanded use of new or more modern electrical appliances, larger electrical distribution systems are required. It is not uncommon to find temporary power taps, extra wires placed under circuit breaker terminals or splices being made in attics or in crawl spaces to add extra electrical equipment to the electrical system. Often, extra fuse or circuit breaker panels are added to existing systems without taking proper steps to ensure that the conductors and equipment supplying the loads are adequate.

In addition, electrical systems are exposed to overvoltage from lighting or power-line crosses, insulation failure in high-to-low voltage transformers, to short circuits and ground faults. In some cases, electrical systems are abused until they fail. The blowing of a fuse or tripping of a circuit breaker often is the first indication of a failure in the electrical system.

Major problems in electrical systems.

There is one major cause of trouble in an electrical distribution system. It is insulation failure. The insulation may be air such as in busways, switchboards and motor control centers where clearance between conductors and metallic electrical equipment is maintained by air space. Rubber, ceramic or thermoplastic insulators provide insulation and separation for conductors as well. Insulation for wire or bus is usually rubber or thermoplastic material.

Insulation failure may result in two kinds of faults, line-to-line or line-to-ground. The least likely failure is between any two conductors of the system, that is, from one phase conductor to another or from one phase conductor to the neutral or grounded conductor. Experience has shown that most insulation failures are line-to-ground faults or from one conductor to the conductor enclosure or equipment.

While the entire National Electrical Code®, and other safety electrical codes, like the Canadian Electrical Code®, are developed and updated to provide protection against electrical fires and shocks, they are not design specifications. However, by following the rules of these codes, we will have an installation that is essentially free from hazard, but not necessarily efficient, convenient, or adequate for good service or future expansion. See Section 90-1.

Insulation resistance.

Previous editions of the Code (e.g., up to the 1965 NEC®) contained recommended values for testing insulation resistance. It was found that those values were incomplete and not sufficiently accurate for use in modern installations, and the recommendation was deleted from the Code. However, basic knowledge of insulation-resistance testing is important.

Measurements of insulation resistance can best be made with a megohm-meter insulation tester. As measured with such an instrument, insulation resistance is the resistance to the flow of direct current (usually at 500 or 1000 volts for systems of 600 volts or less) through or over the surface of the insulation in electrical equipment. The results are in ohms or megohms, but insulation readings will be in the megohm range.

Figure 1-8. 15 kV megohmmeter. Megger® courtesy of Biddle Instruments, Blue Bell, Pennsylvania.

The insulation-resistance test is nondestructive, quite different from a high-voltage or breakdown test. It is made with direct current rather than alternating current and is not a measure of dielectric strength as such. However, insulation resistance tests assist greatly in determining when and where not to apply high voltage.

In general, insulation resistance decreases with increased size of a machine or length of cable, because there is more insulating material in contact with conductors and frame, ground, or sheath.

Insulation-resistance usually increases with higher voltage rating of apparatus because of increased thickness of insulating material.

Insulation-resistance readings are not only quantitative, but are relative or comparative as well, and since they are influenced by moisture, dirt, and

deterioration, they are reliable indicators of the presence of those conditions. Cable and conductor installations present a wide variation of conditions from the point of view of the resistance of the insulation. These conditions result from the many kinds of insulating materials used, the voltage rating or insulation thickness, and the length of the circuit involved in the measurement. Furthermore, such circuits usually extend over great distances and may be subject to wide variations in temperature, which will have an effect on the insulation resistance values obtained. The terminals of cables and conductors will also have an effect on the test values unless they are clean and dry, or guarded.

Records of insulation-resistance should be maintained so failures of conductors or equipment can be predicted. Equipment or conductors can then be repaired or replaced to reduce downtime which often is at inconvenient times causing expensive loss of production.

Differentiate between a "short circuit" and a "ground fault."

It is common practice to call all faults or failures in the electrical system "shorts" or "short circuits." That can lead to misunderstanding, and has done so, because we are not using terms properly. Short circuits and ground faults are different, although they both stem from insulation failure.

To end this confusion and to have a common language for better understanding, we should define the two terms as follows:

Short Circuit

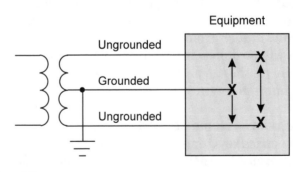

Figure 1-9

Short circuit. "A conducting connection, whether intentional or accidental, between any of the conductors of an electrical system whether it be from line-to-line or from line to the grounded conductor."

In the case of a "short circuit," the failure may be from one phase conductor to another phase conductor or from one phase conductor to the grounded conductor or neutral. For either condition, the maximum value of fault current is dependent on the available capacity the system can deliver to the point of the fault. The maximum value of short-circuit current from line-to-neutral will vary depending upon the distance from the source to the fault. The available short-circuit current is further limited by the impedance of the arc where one is established, plus the impedance of the conductors to the point of short circuit.

Ground Fault

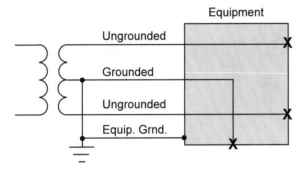

Figure 1-10

Ground fault. "A conducting connection whether intentional or accidental between any of the conductors of an electrical system and the conducting material which encloses the conductors or any conducting material that is grounded or that may become grounded."

In the case of a "ground fault" we have a fault from a hot or ungrounded phase conductor to the conductor enclosure or from the grounded phase conductor to the conductor enclosure (wire-to-conduit, wire-to-motor frame, etc.). It is not common in practice to refer to a conductor that is intentionally grounded, such as a grounded phase or system conductor or equipment grounding conductor, as being in a ground-fault condition.

It is often the case that a fault may easily involve both conditions. The fault may start as a "ground fault" and develop into a "short circuit." The first situation is much more likely than the second, although the second situation can occur.

NEC® requirements.

Article 250 of the National Electrical Code® covers the subject of "Grounding." Grounding is practiced for the protection of electrical installations as well as persons or animals that may come into contact with the electrical system. The Code does not imply here that grounding is the only method that can be used for the protection of electrical installations. Insulation, isolation and guarding are suitable alternatives under certain conditions.

Grounding of specific equipment is covered in several other articles of the National Electrical Code®. Refer to the index of this text, or to the index of the National Electrical Code® or Ferm's Fast Finder Index book for specific sections of the Code that apply to the equipment in question.

Designing electrical systems for safety.

Electrical systems need to be designed to be certain they are adequate for the loads to be served by them. The NEC® provides that electrical services are to be adequate for the calculated load in accordance with Article 220. [Section 230-42(a)] Methods for calculating the minimum capacity for electrical services are found in Article 220 rather than in Article 230. The calculation method for determining the minimum size of the grounded service conductor (often a neutral) is found in Section 220-22.

The NEC® does not require that electrical systems be designed with any additional capacity for future load expansion [Section 90-1(b)]. From a practical standpoint, this should be done. Some designers plan the electrical system for at least 25 percent spare capacity.

Circuit impedance and other characteristics.

Electrical equipment intended to interrupt current at fault levels must be designed and installed so it has an adequate rating for the fault current that is available at its line terminals (Section 110-9). Failure to comply with this important requirement can result in disastrous consequences including destruction of electrical equipment that is itself designed to protect the system. Equipment such as time clocks, motor controllers and lighting contactors that are intended to interrupt current at other than fault levels must be suitable for the current that must be interrupted.

In addition, Section 110-10 requires that, "The overcurrent protective devices, the total impedance, the component short-circuit withstand ratings, and other characteristics of the circuit to be protected shall be selected and coordinated to permit the circuit-protective devices used to clear a fault to do so without extensive damage to the electrical compo-

Interrupting Rating

Circuit Impedance and Other Characteristics

Equipment intended to break current at fault levels shall have an interrupting rating sufficient for the nominal circuit voltage and the current that is available at the line terminals of the equipment.

Equipment intended to break current at other than fault levels shall have an interrupting rating at nominal circuit voltage and current to be interrupted.

Figure 1-11

- Overcurrent protective devices,
- The total impedance,
- The component short-circuit withstand rating,
- Other characteristics of circuit,

Selected and coordinated to permit circuit protective devices to clear fault without extensive damage to the electrical components of the circuit

Fault assumed to be:
- Between two or more of the circuit conductors
- Between any circuit conductor and the grounding conductor or enclosing metal raceway

Figure 1-12

nents of the circuit. This fault shall be assumed to be either between two or more of the circuit conductors, or between any circuit conductor and the grounding conductor or enclosing metal raceway."*

Nearly all components of the electrical system have withstand ratings that must be considered for safety. Electrical system components such as wire, metering equipment, switchboards, panelboards, motor control centers and busways have limited short-circuit ratings.

Overcurrent protective devices such as circuit breakers and fuses should be selected and installed to ensure that the short-circuit withstand rating of components of the system will not be

exceeded should a short circuit or ground fault occur. The overcurrent device that is selected must not only safely interrupt the fault current that is available at its line terminals, it also must limit the amount of energy that is let through to downstream equipment so as to not exceed its rating. See Chapter 11 of this text for an extensive treatment of conductor withstand ratings.

Electric utility companies usually provide information on the short-circuit current (often referred to as fault current) that is available at the terminals of their supply transformer or distribution system. The short-circuit current must then be calculated to the point on the electrical system under consideration. Methods for calculating the available fault current at any point on the electrical system can be found in literature from many circuit breaker or fuse manufacturers. See also the Institute of Electrical and Electronic Engineers (IEEE) "Red Book" (IEEE 141)[1], Underwriters Laboratories product safety standard and "guide card information" for the equipment under consideration[2], the Bussmann *SPD Electrical Protection Handbook*[3] as well as in *Ferm's Fast Finder Index*[4] where, in addition to the point-to-point method, many pages of calculations are presented in an easy-to-use table form.

[1] Available from
IEEE Service Center
445 Hoes Lane, P.O. Box 1331
Piscataway, NJ 08855-1331
(800)678-4333

[2] Available from
Publications Stock
Underwriters Laboratories
333 Pfingsten Road
Northbrook, IL 60062-2096
(847)272-8800 Ext. 42612 or 42622

[3] Available from
Cooper Industries, Bussmann Division
P.O. Box 14460
St. Louis, MO 63178

[4] Available from
International Association of Electrical Inspectors
901 Waterfall Way, Suite 602
Richardson, TX 75080
(800) 786-4234,
Fax (972) 235-3855

Chapter One: The questions included here were developed using material included in this chapter. The answers can be found by reviewing the text. It is also important that students make use of the 1996 NEC®, where many answers can be found. See page 235 for answers.

1. A grounded metal object is forced to take the same zero potential as the ____.
   a. electrode
   b. metal raceway
   c. earth
   d. system

2. Electrical systems are solidly grounded in to limit the voltage to ground during normal operation and to prevent excessive voltages because of ____, line surges or unintentional contact with higher voltage lines.
   a. low voltage
   b. overloads
   c. loose connections
   d. lightning

3. The path to ground from circuits, equipment, and metal enclosures must ____.
   a. be permanent and continuous
   b. have capacity to conduct safely any fault current likely to be imposed on it
   c. have sufficiently low impedance to limit the voltage to ground and to facilitate the operation of the circuit protective devices in the circuit
   d. all of the above

4. Insulation is considered to be air where conductors are mounted on insulators such as on poles, in busways or in equipment, or rubber or ____ material or wire or bus.
   a. listed
   b. approved
   c. acceptable
   d. thermoplastic

5. Measurements of insulation resistance can best be made with a ____ insulation tester.
   a. voltage
   b. megohm-meter
   c. low voltage
   d. high voltage

6. Insulation resistance is the resistance to the flow of direct current (usually at ____ or ____ volts for systems of 600 volts or less) through or over the surface of the insulation in electrical equipment megohm range.
   a. 500 or 1000
   b. 600 or 1500
   c. 800 or 1800
   d. 900 or 2000

7. An insulation-resistance test is made with direct-current rather than alternating-current, and is not a measure of ____ strength as such.
   a. conductor
   b. dielectric
   c. terminal
   d. equipment

8. A conducting connection, whether intentional or accidental, between any of the conductors of an electrical system whether it be from line-to-line or to the grounded conductor defines a ____.
   a. phase fault
   b. short circuit
   c. overload
   d. line disorder

9. A conducting connection, whether intentional or accidental, between any of the conductors of an electrical system and the conducting material which encloses the conductors or any conducting material that is grounded or that may become grounded is defined as a ____.
   a. phase fault
   b. overload
   c. system ground
   d. ground fault

Notes:

Chapter 2

To Ground or Not to Ground

Objectives

After studying this chapter, the reader will be able to understand

- Grounding rules for alternating-current systems of up to 1000 volts.
- Grounding rules for alternating-current systems 1kV and over.
- Which systems cannot be grounded.
- Which systems can be operated ungrounded.
- Use of ground detection systems for ungrounded systems.
- Factors to consider regarding system grounding.

Over the years, there have been great debates over the merits of grounding electrical systems versus leaving them ungrounded. This subject is covered at some length in the Appendix where the history of the development of the grounding rules in effect today are covered.

Many of the decisions about whether or not to ground electrical systems are made for us. The National Electrical Code® requires that electrical systems falling within the parameters of Section 250-5 be grounded. Other electrical systems are permitted to be grounded while some systems are not permitted to be grounded under certain conditions. Some systems today, primarily in the industrial or agricultural sector, are operated ungrounded.

Systems that must be grounded.

In accordance with Section 250-5, the alternating-current systems that must be grounded are as described in detail in Parts A, B, C and D of this text.

Systems Less Than 50 Volts That Must Be Grounded

Over 150 volt supply. Less than 50 volt

Ungrounded supply. Less than 50 volt

Overhead conductors.

Figure 2-1

A. **Less than 50 volt systems:**
 1. Where supplied by transformers if the supply voltage to the transformer exceeds 150 volts to ground.
 2. Where supplied by transformers if the transformer supply system is ungrounded.
 3. Where installed as overhead conductors outside of buildings.

B. **50 to 1000 volt systems supplying premises wiring or premises wiring systems:**
 1. Where the system can be grounded so the maximum voltage to ground on the ungrounded conductors does not exceed 150

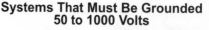

Systems That Must Be Grounded 50 to 1000 Volts

Where the system can be grounded so the maximum voltage to ground does not exceed 150 volts.

3-phase, 4-wire wye systems where neutral is used as circuit conductor.

3-phase, 4-wire delta systems where the midpoint of one phase winding is used as circuit conductor.

Where the grounded service conductor is uninsulated. (Typical for service drops and service-entrance cables.) Sections 230-22, 230-30 and 230-41.

Figure 2-2

volts. Typical systems include:
 - 120 volt, 1-phase,
 - 120/240 volt, 1-phase, and
 - 208Y/120 volt, 3-phase, 4-wire.
 2. Where the system is 3-phase, 4-wire wye connected where the neutral is used as a circuit conductor. Typical systems include:
 - 208Y/120 volt, 3-phase, 4-wire, and
 - 480Y/277 volt, 3-phase, 4-wire.
 3. Where the system is 3-phase, 4-wire delta

connected in which the midpoint of one phase winding is used as a circuit conductor. Typical systems include:

- 120/240 volt, 3-phase, 4-wire, and
- 240/480 volt, 3-phase, 4-wire.

4. Where the grounded service conductor is uninsulated as permitted as follows:

- the grounded conductor of a multiconductor service drop cable (Section 230-22),
- the grounded conductor of an underground service lateral (Section 230-30), and
- the grounded service-entrance conductor (Section 230-41).

Exceptions from grounding.

Five exceptions to the requirement for grounding these alternating-current systems are provided:

1. Electric systems used exclusively to supply industrial electric furnaces for melting, refining, tempering, etc.
2. Separately derived systems used exclusively for rectifiers supplying only adjustable speed industrial drives.
3. Separately derived systems supplied by transformers that have a primary voltage rating less than 1000 volts if all the following conditions are met:
   a. The system is used for only control circuits.
   b. Only qualified persons will service the installation.
   c. Continuity of power is required.
   d. Ground detectors are installed on the control system.
4. Isolated power systems permitted or required in Section 517-61(a)(1) for flammable anesthetizing systems, and ac systems supplying fixed and portable electrical equipments within the cell line working zone. See Section 668-30(a) regarding electrolytic cells used for the production of aluminum, cadmium, chlorine, copper and other products.
5. Three-phase ac systems of 480 to 1000 volts that are high-impedance grounded neutral systems. (An impedance, usually a resistor, limits the ground-fault current of the first fault to a low value.) All the following conditions must be met before high-impedance grounded systems are permitted:

a. Only qualified persons will service the system.
b. Power continuity is required.
c. Ground detectors are installed.
d. Line-to-neutral loads are not supplied.

C. AC Systems of 1 kV and Over.

These ac systems must be grounded if they supply mobile or portable equipment as covered in Section 250-154. Systems supplying portable or mobile high-voltage equipment, other than substations installed on a temporary basis, are required to comply with the rules in (a) through (f) below.

AC Systems of 1 kV and Over Supplying Portable or Mobile Equipment

Eq. Gnd.

- Supplied from system having its neutral grounded through impedance
- Delta-connected system must have neutral derived
- Equipment grounded where system neutral impedance is grounded
- Max. 100 volts from fault current
- Ground-fault detection and relaying
- Grounding electrode isolated

Figure 2-3

(a) Portable or Mobile Equipment. Portable or mobile high-voltage equipment must be supplied from a system having its neutral grounded through an impedance. Where a delta-connected high-voltage system is used to supply portable or mobile equipment, a system neutral shall be derived. This is usually accomplished by means of a zigzag grounding transformer. See Chapter Three of this text for additional information on this subject.

(b) Exposed Noncurrent-Carrying Metal Parts. Exposed noncurrent-carrying metal parts of portable or mobile equipment shall be connected by an equipment grounding conductor to the point at which the system neutral impedance is grounded. This point may be at the service or source of a separately derived system.

(c) Ground-Fault Current. The voltage developed between the portable or mobile equipment frame and ground by the flow of maximum ground-fault current shall not exceed 100 volts. This requirement, no doubt, necessitates an engineering study to determine the voltage drop across the equipment grounding conductor.

(d) Ground-Fault Detection and Relaying. Ground-fault detection and relaying shall be provided to automatically de-energize any high-voltage system component that has developed a ground fault. The continuity of the equipment grounding conductor shall be continuously monitored so as to de-energize automatically the high-voltage feeder to the portable or mobile equipment upon loss of continuity of the equipment grounding conductor.

(e) Isolation. The grounding electrode to which the portable or mobile equipment system neutral impedance is connected shall be isolated from and separated in the ground by at least 20 ft. (6.1 m) from any other system or equipment grounding electrode, and there shall be no direct connection between the grounding electrodes, such as buried pipe, fence, etc.

(f) Trailing Cable and Couplers. High-voltage trailing cable and couplers for interconnection of portable or mobile equipment shall meet the requirements of Part C of Article 400 for cables and Section 710-45 for couplers.

Other ac systems over 1,000 volts are permitted (not required) to be grounded.

This equipment is commonly found in mobile rock crushing plants and batch plants. Other applications are for open pit mining operations. Note that self-propelled mobile surface mining machinery and its attendant electrical trailing cable are no longer covered by the Code. See Section 90-2(b)(2).

D. Separately Derived Premises Wiring Systems.

Electrical systems derived from a generator, transformer, or converter windings that have no direct electrical connection to the supply conductors originating in another system, must be grounded if the system that is derived complies with the conditions in (A), (B), or (C) above. The separately derived system then must be grounded as specified in Section 250-26.

Examples of systems that are separately derived include:
- Transformers with no direct electrical connection between the primary and secondary.

Separately Derived Premises Systems That Must Be Grounded

Transformers with no direct electrical connection between primary and secondary.

Generators with no direct electrical connection with another supply system.

AC or DC systems derived from converter windings.

Figure 2-4

- Generator systems that supply power such as for carnivals, rock crushers or batch plants where the neutral is not connected to the utility system.
- Generator systems used for emergency, required standby or optional standby power that have all conductors including a neutral isolated from the neutral or grounded conductor of another system usually by a transfer switch.
- Ac or dc systems derived from inverters or rectifiers.

Electrical Systems That Are Not Separately Derived Systems

Autotransformer. Not separately derived as the systems have one common conductor.

Not separately derived as the neutral is not switched in the transfer switch.

Figure 2-5

Figure 2-5 illustrates systems that are not separately derived and thus do not fall under the grounding requirements of Section 250-26. These systems include:

- A system supplied by an autotransformer since autotransformers by design have a common conductor.
- Systems from a generator that do not have all conductors, including the neutral conductor, switched by a transfer switch.

The key to determining whether a generator supplied system is a separately derived system is often to examine electrical connections in the transfer switch. If the neutral is switched by the transfer switch, it is a separately derived system and must be grounded in compliance with Section 250-26. If the neutral is not switched, it is not a separately derived system. See Chapter Twelve for additional information on grounding of separately derived systems.

Circuits or systems that cannot be grounded.

Only three systems are not permitted to be grounded by the Code. (See Section 250-7) They are as follows:

Circuits That Are Not Permitted To Be Grounded

Cranes that operate over combustible fibers in Class III locations.

Isolated power systems in Health Care Facilities for hazardous inhalation anesthetizing and wet locations.

Circuits for electrolytic cells as provided for in Article 668.

Figure 2-6

1. Circuits for electric cranes that operate over combustible fibers in Class III locations. (See Sections 250-7(a) and 503-13). This action reduces the likelihood that sparks from faulted equipment will fall onto combustible fibers causing a fire.

2. For health care facilities, those isolated power circuits in hazardous (classified) inhalation anesthetizing locations are required to be supplied by an isolation transformer or other ungrounded source [Section 517-61(a)(1)]. In addition, receptacles and fixed equipment in "wet locations" of hospital patient care areas as defined in Section 517-3 must be protected by ground-fault circuit-interrupter devices where interruption of power to equipment under fault conditions can be tolerated. Where interruption of power under fault conditions cannot be tolerated, protection of these receptacles and fixed equipment can be supplied from isolated, ungrounded sources by ungrounded electrical systems. [Sections 250-7(b) and 517-20(a)]

3. Circuits for electrolytic cells as provided in Article 668. Equipment located or used within the electrolytic cell line working zone or associated with the cell line dc power circuits are not required to comply with Article 250. See Sections 250-7(c) and 668-3(c) Exception No. 3.

Systems permitted to be operated ungrounded.

At times, the plant owner or engineer will choose to operate electrical systems ungrounded. These systems usually are in industrial or agricultural applications and usually are 480-volt, three-phase, three-wire systems. Where ungrounded systems are installed, an effort often is being made to obtain an additional degree of service continuity.

Systems Permitted to be Ungrounded

480 volt, 3-phase, 3-wire
2300 volt, 3-phase, 3-wire
4600 volt, 3-phase, 3-wire
13.8 kV, 3-phase, 3-wire

Figure 2-7

Typical systems that are operated ungrounded include:
- 480 volt, 3-phase, 3-wire, delta connected.
- 2,300 volt, 3-phase, 3-wire, delta connected.
- 4,600 volt, 3-phase, 3-wire, delta connected.
- 13,800 volt, 3-phase, 3-wire, delta connected.

First Ground Fault on Ungrounded System

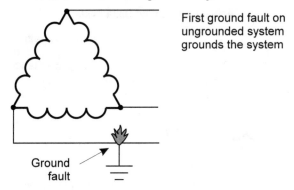

First ground fault on ungrounded system grounds the system

Ground fault

Figure 2-8

Since the system is ungrounded, the occurrence of the first ground fault (as distinguished from a short circuit) on the system will not cause an overcurrent protective device such as for the service or feeder to open. This fault does, however, ground the system. In essence, this system becomes a corner grounded delta system. Little, if any, current will flow in this first ground fault. See Chapter Three of this text for additional information on grounding of these ungrounded systems.

Second Ground Fault on Ungrounded System

High-impedance path through metal enclosures, loose raceway connections, etc.

Ground fault

Ground fault

Figure 2-9

A second ground fault from the same system that originally was ungrounded becomes a phase-to-phase fault at the service, a feeder or circuit. This will usually cause one or more overcurrent protective devices to open provided adequate current flows in this path to cause the circuit protective devices to operate. A major concern for this type of system happens where the first and second faults may be located some distance apart. Often, these faults may occur from line-to-conduit or metallic enclosure such as wireway, pull boxes, busway or motor terminal housings in different parts of the plant. Where this occurs, often, a high-impedance path for current flow is established. In some cases, it has been found that a great deal of heat along with arcing and sparking is produced in this fault path due to loose connections. Every conduit coupling and locknut connection to enclosures in the fault-current path must be tight to provide an adequate and low-impedance path.

It is important for safety reasons, as well as for system continuity, that maintenance personnel locate and eliminate ground faults on ungrounded systems as soon as practical and especially before the second ground fault occurs on the system.

Ground-detector systems.

Commercially manufactured ground-detection equipment is available. This equipment, which can be located at the service equipment or in feeder distribution panels, can be set to operate an overcurrent relay or shunt-trip circuit breaker or to operate a visual or audible signalling system to indicate a ground-fault condition.

This monitoring equipment is not required by the electrical code but is a good practice from a plant management perspective. Successful operation of an ungrounded electrical system depends on good system maintenance and prompt elimination of the first ground fault.

Sophisticated equipment is now available to identify the part of the electrical system where the fault is located while the system is energized. This significantly speeds detection and repair.

Ground detector lights.

Often, ground detector lights or a neutralizer or potentializer plug are installed to indicate that a ground fault has occurred on the ungrounded system. The 7½ watt indicator lights are connected to the lines through 18,000 ohm resistors. A tap is made to each resistor to give 120 volts to the lamp.

The lamp burns until its phase goes to ground, at which time there is no or little potential across the lamp, and it stops glowing, thus identifying the faulted phase.

Ground Detector Lights

Figure 2-10

This indication is intended to alert the maintenance personnel to the problem so the ground-fault can be corrected during hours when the plant is not operating. The plant can continue to operate with one-phase grounded thus preventing costly production down times. In some cases, down-time in production plants can cost thousands of dollars per minute.

A wide variety of "homemade" systems has been installed over the years, some of which are downright dangerous. Some of these have consisted of nothing more than two "barn socket" lampholders with 240-volt lamps that are connected phase-to-ground.

Systems and circuits.

The grounding of systems and circuits is intended to limit the voltage that may be impressed on the circuit through exposure to lightning or other voltages which may be higher than that for which the circuit is designed. Such grounding also will limit the maximum voltage to ground under normal voltage conditions.

Ungrounded system problems.

An ungrounded system exists only in theory, in a laboratory or at the electrical distribution transformers hanging on the pole before connection to the plant electrical system. In the real world, ungrounded systems are grounded through distributed leakage capacitance of the system.

Actual Ungrounded Delta System

Ungrounded delta system is grounded through capacitance.

Distributed leakage capacitance

Figure 2-11

When any conductor is installed in close proximity to grounded metal, there is a capacitance between them that is increased as the distance between the conductors is reduced. In 600-volt systems, the two greatest sources of capacitance to ground are conductors in metal conduit and motor windings. In both cases, conductors are separated from grounded metal by fairly thin insulation. The capacitance, to ground is known as the leakage capacitance and the current flowing from the conductors to ground is known as the leakage current or charging current. This capacitance is distributed throughout the electrical system but acts like it is a single, lumped capacitance.

Disadvantages of operating systems ungrounded include:

1. Power system overvoltages are not controlled. In some cases, these overvoltages are passed through transformers into the premises wiring system. Some common sources of overvoltages include: lightning, switching surges and contact with a high-voltage system.
2. Transient overvoltages are not controlled, resulting in insulation degradation and failure.
3. System voltages above ground are not controlled.
4. Destructive arcing burndowns if a second fault occurs before the first fault is cleared.

Ungrounded systems have the characteristic that they are subject to relatively severe transient overvoltages. Such overvoltages may be owing to external disturbances as well as to internal faults in the wiring system and easily can reach a value of five to six times normal voltage. An actual case involved a 480-volt ungrounded system. Line-to-ground potentials in excess of 1200 volts were measured on a test meter. The source of the trouble was traced to an intermittent or sputtering line-to-ground fault in a motor starting autotransformer. These faults are not uncommon on 480-volt ungrounded systems. During the two-hour period that this arcing fault existed, between 40 and 50 motor windings had failed.

Circuit-switching operations can also be responsible for the creation of transient overvoltages in ungrounded systems. These generally are of short duration and are not in excess of two to three times normal system voltage.

Experience has shown that these overvoltages, that easily reach several times the system voltage, may cause failure of insulation at locations on the system other than at the point of the fault and may result in system failures. This often occurs at a system weak point such as in a motor winding.

Locating the ground fault can be troublesome. While it is easy to spot a ground fault on a one-line diagram, locating it in a plant with a complex electrical system can be much more difficult unless sophisticated ground-fault detection equipment has been installed. The first step is to open the feeders one at a time and observe the ground detection indicator. After finding the feeder with the ground fault, branch circuits are disconnected, one at a time, until the offending circuit is located. A significant loss of plant operation time can occur during this process. This is contrasted with a grounded system where only the offending equipment is taken off the line by the circuit protective devices.

Often, overcurrent devices are set above the current level of the fault in ungrounded systems. When arcing faults occur, destructive burn-downs result. The fault arc releases a tremendous amount of energy such that conductors and metal enclosures in the vicinity are destroyed.

When the first ground fault occurs on the 480-volt system, the other conductors of the system rise to a level of 480 volts to ground. This presents an additional risk of shock to operation and maintenance staff. This can be contrasted to a 480Y/277-volt grounded wye system where the voltage to ground does not exceed 277 volts while the phase-to-phase voltage is 480 volts.

Factors to consider regarding system grounding.

Where grounding of the electrical system is optional, the pros and cons of grounding must be carefully weighed by the plant owner or electrical designer so as to make the best decision.

In the long run, greater service continuity may be obtained with grounded systems rather than ungrounded ones. Faults can be isolated to the feeder or circuit affected and cleared without disrupting the entire system. This is obviously a major consideration if the equipment or circuit affected is critical to the plant operation.

High-impedance grounded systems should be considered where the plant operation cannot tolerate electrical system disruption from the first ground fault. This system has all the advantages of an ungrounded system, so far as operation of the plant or system with one phase faulted to ground is concerned, with none of the disadvantages of an ungrounded system. While the initial cost of the system can be fairly significant, it can pay for itself many times over the installation cost by operational savings in more reliable plant operation. See Chapter Four of this text for additional information on high-impedance grounded neutral systems.

Bolted faults.

A common myth is that ground-faults are "bolted" or solidly connected and that a great deal of fault-current will flow to cause the overcurrent device to clear the fault. Bolted faults rarely occur, while sparking, intermittent or arcing faults are quite common.

Arcing faults produce a great deal of heat in the vicinity of the fault and can lead to destructive burn-downs of electrical switchboards and motor control centers. This is the reason the Code requires equipment ground-fault protection systems for certain voltage and amperage services and feeders. See Chapter Fifteen of this text for additional information on this subject.

For the purposes of determining adequate interrupting ratings of overcurrent protective devices as well as bracing of bus bars, etc., bolted fault-current values must be used. This is considered to be the "worst case" condition.

Chapter Two: The questions included here were developed using material included in this chapter. The answers can be found by reviewing the text. It is also important that students make use of the 1996 NEC®, where many answers can be found. See page 235 for answers.

1. Where operating at less than ____ volts, alternating-current systems are required to be grounded where supplied by transformers if the supply voltage to the transformer exceeds ____ volts to ground.
   a. 80 - 100
   b. 70 - 110
   c. 60 - 140
   d. 50 - 150

2. Where operating at less than 50 volts, ac systems are required to be grounded where supplied by transformers if the transformer supply system is ____.
   a. bonded
   b. ungrounded
   c. identified
   d. approved

3. Conductors installed on the outside of buildings where ac systems operate at less than 50 volts are required to be grounded when they are run as ____ conductors.
   a. overhead
   b. underground
   c. optical fiber
   d. Type IGS

4. Alternating-current system grounding is required for 50 to 1000 volt systems supplying premises wiring or premises wiring systems where the system can be grounded so the maximum voltage to ground on the ungrounded conductors does not exceed ____ volts.
   a. 180
   b. 150
   c. 240
   d. 208

5. Alternating-current system grounding is required for 50 to 1000 volt systems supplying premises wiring or premises wiring systems where the system is 3-phase, 4-wire, wye connected in which the neutral is used as a ____ conductor.
   a. bonding
   b. circuit
   c. equipment grounding
   d. isolated

6. Alternating-current system grounding is required for 50 to 1000 volt systems supplying premises wiring, or premises wiring systems where the system operates at 3 phase, 4 wire, delta connected in which the midpoint of one phase is used as a ____.
   a. bonding jumper
   b. equipment grounding conductor
   c. circuit conductor
   d. switch leg

7. Alternating-current system grounding is required for 50 to 1000 volt systems supplying premises wiring, or premises wiring systems where the grounded service conductor of a multiconductor service drop cable is ____.
   a. green
   b. insulated
   c. bonded
   d. uninsulated

8. Alternating-current system grounding is required for 50 to 1000 volt systems supplying premises wiring, or premises wiring systems where the grounded service conductor of an underground service lateral is ____.
   a. approved
   b. uninsulated
   c. insulated
   d. bonded

9. Alternating-current system grounding is required for 50 to 1000 volt systems supplying premises wiring, or premises wiring systems where the grounded service entrance conductor is ____.
   a. identified
   b. approved
   c. uninsulated
   d. insulated

10. Alternating-current system grounding is not required for 50 to 1000 volt electric systems used exclusively to supply industrial electric furnaces for ____.
   a. melting
   b. refining
   c. tempering
   d. all of the above

11. Alternating-current system grounding is not required for 50 to 1000 volt separately derived systems used exclusively for rectifiers supplying only ____.
   a. motor control centers
   b. adjustable speed industrial drives
   c. commercial buildings
   d. class III hazardous (classified) locations

12. Alternating-current system grounding is not required for 50 to 1000 volt separately derived systems supplied by transformers that have a primary voltage rating less than 1000 volts if which of the following conditions is or are met. ____
   a. system is used for only control circuits.
   b. only qualified persons service installation, ground detectors are installed on the control system.

c. continuity of power is required.

d. all of the above

13. Alternating-current system grounding is not required for 50 to 1000 volt systems supplying premises wiring, or premises wiring systems where ____ that are permitted or required in for flammable anesthetizing systems in health care facilities, and ac systems supplying fixed and portable electrical equipments within an electrolytic cell line working zone.

a. grounded power systems

b. isolated power systems

c. impedance grounded systems

d. medium voltage power systems

14. Alternating-current system grounding is not required for 50 to 1000 volt systems supplying premises wiring, or premises wiring systems where three-phase ac systems of 480 to 1000 volts that are high-impedance grounded neutral systems if which one of the following conditions is or are met ____:

a. only qualified persons will service the system.

b. power continuity is required.

c. ground detectors are installed, line-to-neutral loads are not supplied.

d. all of the above

15. Ac systems of ____ and over must be grounded if they supply mobile or portable equipment as covered in Article 710.

a. 250 volts

b. 1kV

c. 480 volts

d. 600 volts

16. Ac systems operating at over ____ volts are permitted (not required) to be grounded where they do not supply mobile or portable equipment.

a. 1000

b. 300

c. 277

d. 480

17. Equipment operating at over ____ volts is commonly found in mobile rock crushing plants and batch plants. Other applications are for open pit mining operations.

a. 277

b. 300

c. 600

d. 1000

18. A ____ premises wiring system is one that is derived from a generator, transformer or converter windings that have no direct electrical connection to the supply conductors originating in another system.

a. identified

b. open neutral

c. isolated power

d. separately derived

19. Separately derived systems must be grounded as specified in Section ____.

a. 250-26

b. 250-72

c. 250-114

d. 250-80

20. Examples of separately derived systems include ____ with no direct electrical connection between the primary and secondary.

a. phase converters

b. transformers

c. elevator motors

d. Class 1 systems

21. Examples of separately derived systems include generator systems used for emergency, required standby or optional standby power that have all conductors including a neutral switched by a ____.

a. double pole

b. single pole

c. transfer switch

d. backfed device

22. Examples of separately derived systems include ac or dc systems derived from ____.

a. inverters

b. rectifiers

c. generators

d. all of the above

23. Certain systems are permitted to be operated ungrounded and usually are located in industrial or agricultural applications. Typical systems that are operated ungrounded include which of the following ____.

a. 2,300 volt, three-phase, three-wire, delta connected.

b. 4,600 volt, three-phase, three wire, delta connected.

c. 13,800 volt, three phase, three-wire, delta connected.

d. all of the above

24. Often, ground detector lights or a potentializer plug are installed to indicate that a ground fault has occurred on the ____ system.

a. ungrounded

b. grounded

c. bonded

d. identified

Chapter **3**

Grounding Electrical Systems

Objectives

After studying this chapter, the reader will be able to understand

- Rules for system conductor to be grounded.
- Proper identification of grounded conductor.
- Methods of grounding electrical systems.
- Delta bank grounding.
- Grounding rules for ungrounded systems.
- Corner grounded delta systems.

Definitions.

Ground: "A conducting connection, whether intentional or accidental, between an electrical circuit or equipment and the earth, or to some conducting body that serves in the place of the earth."*

Grounded: "Connected to earth or to some conducting body that serves in place of the earth."*

Grounded Conductor: "A system or circuit conductor that is intentionally grounded."*

The electrical system.

As covered in Chapter 2 of this text, Section 250-5 of the NEC® requires that many electrical systems be grounded. It is important that we understand that we are dealing with the electrical system and not the service equipment or disconnecting means at this point.

Electrical System

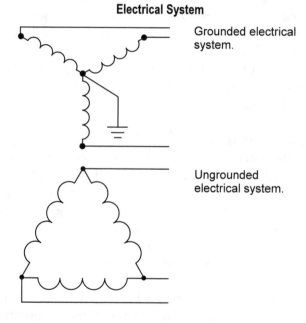

Grounded electrical system.

Ungrounded electrical system.

Figure 3-1

The electrical system is delivered to the customer by the serving utility either grounded or ungrounded. Electrical utilities have tariffs and standards that dictate whether or not they will deliver a system of a certain voltage level and phase configuration as grounded or ungrounded systems. Many power companies insist that all low voltage (600 volts and under) systems be grounded. Others will supply 3-phase, 480 volt delta connected systems ungrounded while they insist on furnishing 480Y/277 volt systems in only a grounded wye configuration.

Large industrial plants may purchase power at medium or high voltage levels such as 69,000 volt, 115,000 volt, or 230,000 volt levels and own and maintain their electrical distribution systems. Transformers, capacitor banks, controls, overcurrent devices and relaying systems are then installed at customer-owned switchyards. Power is distributed to utilization points on the premises where transformers are installed as needed for the voltages desired. Generally, electrical systems at the utilization level are grounded at the voltage levels and configurations required by the NEC®.

Conductor to be grounded.

In Section 250-25, the Code specifies which conductor in an alternating current system shall be grounded. For ac premises wiring systems, the conductor to be grounded shall be as specified in 1 through 5 below.

1. Single-phase, 2-wire: one conductor.
2. Single-phase, 3-wire: the neutral conductor.
3. Multiphase systems having one wire common to all phases: the common conductor.
4. Multiphase systems requiring one grounded phase: one phase conductor.
5. Multiphase systems in which one phase is used as in (2) above: the neutral conductor.

Identification of grounded conductor.

Grounded conductors are required to be identified by the means specified in Sections 200-6 and 310-12. No. 6 and smaller insulated grounded conductors are required to be identified by a continuous white or natural gray outer finish along their entire length. Four exceptions to this requirement are provided in Section 200-6. For our discussion here, Exception No. 3 is most significant. For installations that will be serviced by qualified persons only, the grounded conductors of multiconductor cables that are other than white or natural gray are permitted to be marked at their termination by a distinctive white marking or other equally effective means to identify them as grounded conductors.

For aerial cable, the means of identifying the grounded conductor can be by marking the conductor or by employing a ridge on the exterior of the grounded conductor.

Conductor To Be Grounded

1. Single-phase; 2-wire:
one conductor
(either one).

2. Single-phase, 3-wire:
the neutral conductor.

3. Multiphase system
one common wire:
the common conductor.

4. Multiphase systems
requiring one grounded
phase:
one phase conductor.

5. Multiphase systems in
which the midpoint of
one transformer winding
is used as a neutral
conductor:
the neutral conductor.

Figure 3-2

Identification of Grounded Conductor

No. 6 and smaller.
• **Continuous white or natural gray.**
• **Qualified persons service installation -
identify multiconductor cables at
terminations.**

Larger than No. 6.
• **Continuous white or natural gray.**
• **Distinctive white marking at terminations.**
• **Qualified persons service installation -
identify multiconductor cables at
terminations.**

Conductors larger than No. 6 are permitted to be identified either like conductors No. 6 and smaller, or at the time of installation, by distinctive white marking at terminations. Where only qualified persons will service and maintain the installation, grounded conductors of multiconductor cables are permitted to be permanently identified at terminations by a distinctive white marking or by other effective means.

Methods of grounding.

There are a variety of methods used to ground electrical systems. The method chosen will vary, depending upon the system voltage, code requirements, plant owner specifications or engineer philosophy. The various methods commonly used are shown in Figure 3-4 and are as follows:

(1) **Solidly grounded:** no intentional grounding impedance.

Solidly grounding refers to the connection of the neutral of a generator, power transformer, or grounding transformer directly to the station ground or earth without intentionally introducing an impedance.

(2) **Effectively grounded:** to permit the use of reduced rated (80 percent) surge arresters.

Grounding of surge arresters rated near, but not less than, 80 percent of line-to-line voltage constitutes an effectively grounded system. This will carry with it a line-to-ground circuit current of at least 60 percent of the three-phase short-circuit value.

(3) **Reactance grounded:** the system is grounded by the system neutral through a reactor. This grounding is accomplished by a reactor which is a device that introduces impedance to the circuit, the principal element of which is reactance.

Methods of System Grounding

1. Solidly grounded. No intentional grounding impedance.

2. Effectively grounded. Grounded through surge arresters.

3. Reactance grounded. Grounded through an inductor.

4. Resistance grounded. Grounded through a resistor. Typically 50 to 600 amperes ground-fault current.

5. High-Resistance grounded. Grounded through the highest permissible resistance. Typically 1 to 10 amperes ground-fault current.

6. Ungrounded. No intentional system grounding connection.

Figure 3-3

Ungrounded systems are used in industrial plants and where desired for manufacturing processes to give an additional degree of service continuity. While a ground-fault on one phase of an ungrounded system generally does not cause a service interruption, the occurrence of a second ground fault on a different phase before the first fault is cleared does result in an outage. A ground detection system is recommended, and adequate maintenance and repair will minimize interruptions.

Low-voltage systems 600 volts and below are almost always solidly grounded; medium-voltage systems are usually either solidly or resistance grounded; and high-voltage systems above 34.5 kV are nearly always effectively grounded or ungrounded.

(4) **Resistance grounded:** intentional insertion of resistance into the system grounding connection.

For resistance grounding, the system is grounded by connecting the system neutral to ground through a resistor. This grounding is accomplished by a resistor which introduces impedance to the circuit, the principal element of which is resistance.

(5) **High-resistance grounded:** the insertion of nearly the highest permissible resistance into the grounding connection.

This grounding is accomplished by a resistor which introduces impedance to the circuit, the principal element of which is resistance. For high-resistance grounding, the system is grounded by the system neutral through a resistor (may be a grounding transformer) up to 10-ampere level. High-resistance grounding maintains control of transient overvoltages, but may not furnish sufficient current for ground-fault relaying. The protective scheme usually associated with high-resistance grounding is detection and alarm rather than immediate tripout. See Section 250-27 of the NEC® and Chapter 4 of this text for additional information.

(6) **Ungrounded:** no intentional system grounding connection.

Delta bank grounding.

Where three transformers are connected in a delta bank, only one transformer may have its midpoint grounded. Any attempt to use a second transformer of the delta bank to supply a second 3-wire, single-phase source would require grounding the midpoint of a second transformer. However, since there will be a difference of potential between the midpoints of the different transformers, there would be an abnormal flow of current through both

240 Delta Bank Voltages

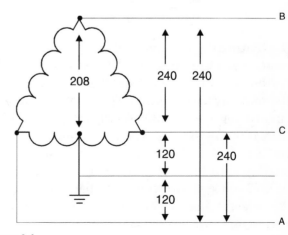

Figure 3-4

grounded neutrals. This would sooner or later show up in heating or a short circuit, depending on how both neutrals were connected.

With midpoint grounding of one phase in a delta bank, one of the phase conductors will operate at a higher voltage to ground than the other two. In practice this "high leg" is identified as required by Sections 215-8, 230-56 and 384-3(e) by orange color coding or other effective means.

The voltage of the leg with the higher voltage to ground is determined by the following formula: ½ of phase-to-phase voltage x 1.732 = high leg voltage to ground. For example, a 240 volt, 3-phase system center tapped to establish a neutral: ½ of 240 = 120 x 1.732 = 208V high leg voltage to ground.

Grounding of ungrounded systems.

In some cases, it is desirable to ground electrical systems that originally were installed ungrounded. See Chapter Two of this text for a thorough discussion of the subject. There are four methods commonly in use for grounding of ungrounded systems of 600 volts or less. Solid grounding is used in all the methods. They are as follows:

Grounding Wye-Connected Bank

Grounded by connecting midpoint to grounding electrode.

Common voltages:
- 208Y/120
- 480Y/277
- 575Y/332
- 600Y/346

Figure 3-5

(1) Grounding the neutral of wye-connected secondary windings of a transformer bank.

As shown in Figure 3-5, this is the most universal and commonly used method of grounding a system. Standard voltages are 208Y/120, 480Y/277, 575Y/332 and 600Y/346 volts. The first two voltage systems, 208Y/120 and 480Y/277, are in most common use today with the growth tending in favor of the 480Y/277 volt system owing to the better economies of that system. In general, the primary windings of the transformers serving those systems are delta connected.

(2) Grounding a delta bank with a zigzag grounding transformer.

This method is best adapted to an existing 3-wire, 3-phase delta-connected distribution system which is ungrounded and where it is desired to ground the system to obtain the advantages gained through operating the system grounded. For new systems, the use of transformers with wye-connected secondaries are advisable to obtain a grounded system. The neutral derived by the zigzag transformer can be used as a system current carrying conductor if the zigzag grounding transformers are sized for the maximum unbalanced current. See Section 220-22.

Grounding Delta Bank with Zigzag Transformer

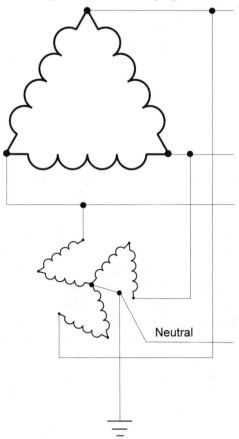

Neutral

Figure 3-6

A zigzag transformer, shown applied to a delta system in Figure 3-6 obtains its name from the manner in which the windings are installed and connected. Windings for each phase are on the same core leg. All windings have the same number of turns but each pair of windings on the same core leg are wound in opposite directions. The impedance of the transformer to 3-phase currents is so high that under normal conditions only a small magnetizing current will flow. If a ground fault develops on one phase of the system, the transformer impedance to ground current is so low that a high ground current will flow. The ground current will divide into three equal parts in the three phases of the grounding transformer.

Such a 3-phase zigzag transformer has no secondary winding. A grounding transformer is required to carry rated current for only a short time (the duration of the ground fault). The short time kVA rating of such a grounding transformer may thus be equal to the rating of a regular 3-phase transformer yet be only about 10 percent of the physical size.

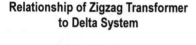

Relationship of Zigzag Transformer to Delta System

Figure 3-7

Electrically, a zigzag grounding transformer connection appears to superimpose a wye connected system inside the delta connected transformers. This wye-type connection does permit the transformer to be used like a wye transformer with phase-to-neutral loads to be utilized. In this case, it is important for the zigzag transformer to be sized for the calculated or connected load.

A wye-delta grounding transformer may also be used to provide a neutral for an existing delta-connected ungrounded system, but the use of the zigzag grounding transformer is more practical and economical.

The grounding transformer may be connected to the system, as shown in Figure 3-6, directly on the power transformer secondaries. Where a grounding transformer is so connected, one grounding transformer is required for each delta-connected power transformer bank. In this manner, the switching out of any one transformer bank will not

disturb the secondary system ground. The grounding transformer and the power transformer are considered a single unit, both being protected by the overcurrent protective means provided for the main power transformer.

The grounding transformer also may be connected directly to the main bus through its own overcurrent protection means. In that event, an alarm should be provided to indicate the system is operating ungrounded if the grounding transformer should be disconnected from the line.

Where a grounding transformer is used on low-voltage systems (600 volts and below), it is important that the equipment grounding conductor be connected to the neutral of the grounding transformer in such a way as to provide a low-impedance path for ground-fault currents to flow back to the system. The same grounding electrode should be used for the neutral of the grounding transformer and the equipment grounding conductor and a common grounding conductor used for both the equipment grounding conductor and the neutral.

The same disconnecting means and overcurrent protection means for the system would be used as described under method (1).

Grounding Delta Transformer Banks

Grounding midpoint of one transformer. (If three, one-phase transformers.)

Figure 3-8

(3) Grounding the midpoint of one of the transformers in a delta-connected system.

This method of grounding is commonly used, especially in smaller distribution systems, to provide a three-phase power source and a three-wire single-phase lighting source from the same transformer bank. This is a three-phase, four-wire delta-connected system. At the same time, the advantages of a neutral grounded system are obtained. It is common for the serving utility to use three single-phase transformers connected in a delta configuration for this system. In such a system, the single-phase transformer which supplies the lighting load usually is sized larger than the other two single-phase transformers which supply the power load only.

Since the conductors supplying the 3-phase power will now be from a grounded source, the grounded conductor must be run to the service equipment. There it must be connected to the equipment grounding conductor of the power system and the grounding electrode. This provides a ground-fault current return path of low impedance. Such a connection enhances the safety of the power system. If the power service and the lighting service go to the same building, which is usually the case, the grounding electrode conductor from the power service, as well as the grounding electrode conductor from the lighting service, must be connected to the same grounding electrode. See Chapter Four of this text for additional information on the subject.

(4) Grounding One Corner of a Delta System.

In the past, most 3-phase ungrounded delta distribution systems comprised three single-phase transformers which were connected delta-delta. The main reason was to be able to continue operating if one transformer failed by disconnecting the faulted transformer and operating the bank open-delta, although at reduced capacity.

When the advantages of grounding became very apparent, it first was the practice to ground one corner of the delta. This is shown in Figure 3-9. The grounded leg must be positively identified white or natural gray throughout the system. (Section 200-6)

Where the grounded conductor is disconnected by the switch or circuit breaker, it is important that a bonding connection be made on the line side of the disconnect or circuit breaker. If this is not done, the electrical equipment will become electrically isolated from the grounded conductor when the switch or circuit breaker is in the open position. In this case, any fault that occurs would cause the enclosure to raise to a hazardous potential above ground equal to the system voltage.

Generally, fuses are not permitted to be installed in the grounded conductor. Section 230-90(b) does not allow an overcurrent device, other than a circuit breaker that opens all conductors, to be inserted in the grounded service conductor. Section 240-22 generally prohibits connecting an

Fusible Switch for Corner Grounded Delta System

Corner grounded delta system

Grounded service conductor

Service equipment

Motor Controller

Equipment grounding conductor

Motor

Figure 3-9

overcurrent device in series with any conductor that is intentionally grounded. Exception No. 1 permits an overcurrent device to be used that opens all conductors of the circuit, including the grounded conductor, and is so designed that no pole can be operated independently. This requirement means that three-pole circuit breakers can be used for this purpose while a three-pole switch with fuses could not as the switch is not the overcurrent device, the fuse is.

Section 430-36 permits a fuse to be inserted in the grounded conductor for the purpose of providing motor overload protection. However, based on the examination of Code requirements discussed in the previous paragraph, this application would be limited to being located downstream from the service equipment.

Keep in mind that Section 250-23(b) requires that the grounded system conductor be run to each service disconnecting means and be bonded to the service disconnecting means enclosure. This is usually accomplished by connecting it to a terminal bar that is mounted inside the service enclosure. Nothing in the Code permits this bonding connection to be interrupted by a switch or circuit breaker.

Corner Grounded Delta System

Figure 3-10

From the service on throughout the system, the grounded conductor must be insulated from equipment and enclosures. [Section 250-23(a)] An equipment grounding conductor is carried with the circuit to ground equipment and enclosures such as disconnecting means, motor controllers, and other noncurrent-carrying equipment that are required to be grounded.

Where it is desirable to disconnect the grounded system conductor from the feeder, it is acceptable to route the conductor from the terminal bar through the switch or circuit breaker. It is not uncommon to see three-pole switches with a copper bar or other suitable slug inserted in the grounded leg.

For purposes of installing or grounding a corner grounded delta system, it is helpful to think of it as a single-phase system. This is illustrated in Figure 3-10. The system is grounded at one corner of the delta. Three conductors are taken to the service where the two ungrounded conductors connect to the disconnecting means or circuit breaker. The grounded service conductor is connected to the neutral terminal bar where it is bonded to the enclosure and connected to the grounding electrode conductor.

Where circuit breakers are used as the disconnecting means for corner grounded systems, they must be marked with a voltage rating suitable for the system voltage and will have a "straight voltage rating" such as 240V or 480V. Two-pole circuit breakers that are suitable for a corner grounded delta system are marked "1-phase/3-phase." (Section 240-85)

Chapter Three: The questions included here were developed using material included in this chapter. The answers can be found by reviewing the text. It is also important that students make use of the 1996 NEC®, where many answers can be found. See pages 235-236 for answers.

1. A conducting connection, whether intentional or accidental, between an electrical circuit or equipment and the earth, or to some conducting body that serves in the place of the earth is defined as a ____.
   a. grounded conductor
   b. ground
   c. grounding electrode
   d. bonded conductor

2. Connected to earth or to some conducting body that serves in place of the earth is defined as ____.
   a. grounded
   b. a grounded conductor
   c. being identified
   d. being bonded

3. A system or circuit conductor that is intentionally grounded is defined as one that is ____.
   a. identified
   b. bonded
   c. a grounded conductor
   d. a grounding conductor

4. No. ____ and smaller insulated grounded conductors are required to be identified by a continuous white or natural gray outer finish.
   a. 6
   b. 4
   c. 2
   d. 1

5. For installations that will be serviced by qualified persons only, grounded conductors of ____ cables are permitted to be marked at their terminations.
   a. Type IGS
   b. Type TFFN
   c. multiconductor
   d. single

6. Conductors ____ than No. 6 are permitted to be identified either like conductors No. 6 and smaller, or at the time of installation, by distinctive white marking at their terminations.
   a. of aluminum or smaller
   b. of copper or smaller
   c. smaller
   d. larger

7. In practice, a "high leg" is identified as required by Sections 215-8, 230-56 and 384-3(e), by a ____ color coding or other effective means.
   a. green
   b. brown
   c. orange
   d. yellow

8. Grounding of surge arresters rated near, but not less than, ____ percent of line-to-line voltage constitutes an effectively grounded system.
   a. 70
   b. 60
   c. 50
   d. 80

9. For resistance grounding, the system is grounded by connecting the system neutral to ground through a ____.
   a. resistor
   b. terminal
   c. ground rod
   d. generator

10. The insertion of nearly the highest permissible resistance into the grounding connection results in a system that is ____.
   a. extra low resistance grounded
   b. medium resistance grounded
   c. low resistance grounded
   d. high resistance grounded

11. High-resistance grounding maintains control of transient overvoltages, but may not furnish sufficient current for ground-fault ____.
   a. detection
   b. relaying
   c. alarms
   d. conditions

12. Low-voltage systems operating at 600 volts and below are almost always solidly grounded; medium-voltage systems are usually either solidly or resistance grounded; and high-voltage systems above ____ kV are nearly always effectively grounded or ungrounded.
   a. 12.7
   b. 10.4
   c. 15.3
   d. 34.5

13. A wye-delta grounding transformer may also be used to provide a neutral for an existing delta-connected ungrounded system, but the use of the ____ grounding transformer is considered as being more practical and economical.
   a. converter
   b. autotransformer
   c. zigzag
   d. Scott

14. Where a grounding transformer is used on low-voltage systems operating at 600 volts and below, it is important that the equipment grounding conductor be connected to the neutral of the grounding transformer in such a way as to provide a ____ path for ground-fault currents to flow back to the system.
    a. high-impedance
    b. low-impedance
    c. current limiting
    d. straight

15. The same grounding electrode should be used for the ____ of the grounding transformer and the equipment grounding conductor and a common grounding conductor used for both the equipment grounding conductor and the neutral.
    a. unidentified conductor
    b. equipment grounding conductor
    c. bonding jumper
    d. neutral

16. In the past, most 3-phase ungrounded delta distribution systems comprised three single-phase transformers that were connected ____.
    a. wye-delta
    b. delta-wye
    c. delta-delta
    d. delta

17. For purposes of installing or grounding a corner grounded delta system, it is best to think of it as a ____ system.
    a. three phase
    b. single phase
    c. 5-wire
    d. 4-wire

Grounding Electrical Services

Objectives

After studying this chapter, the reader will be able to understand.

- Important elements for grounding of electrical services.
- Proper location of service grounding connection.
- Rules for low-impedance grounding electrode connections.
- Dwelling unit services and feeders.
- Proper sizing of grounded service conductor.
- Rules for parallel service conductors.
- Rules for multiple services to one building.
- Rules for high-impedance grounded systems.
- Grounding requirements for instrument transformers, relays, etc.
- Hazards of operating an ungrounded service from a grounded system.

General.

As discussed in Chapter Two, electrical systems are furnished to the premises by the serving utility as either grounded or ungrounded. At the service disconnecting means, the service is either grounded, left ungrounded or may be resistance grounded. How the services are treated regarding grounding depends on the type of system installed, design criteria and Code rules.

Grounded electrical systems.

One of the most important elements of a grounding system is contained in Section 250-23(b). This section requires that:

Grounded Conductor Run to Each Service Disconnecting Means Enclosure

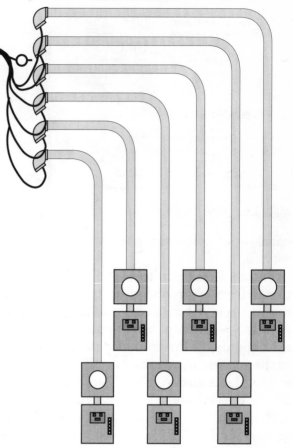

Figure 4-2

Grounded Electrical System

1. Grounded conductor must be run to each service disconnecting means.

2. It must be bonded to the service disconnecting means enclosure.

3. It must be routed with the phase conductors.

4. It must be not smaller than the grounding electrode conductor.

5. For conductors larger than Table 250-94, must be not less than 12½ percent of area.

6. Where service conductors are installed in parallel, base on size of parallel service conductors.

Figure 4-1

1. Where an ac system operating at less than 1000 volts is grounded at any point, the grounded conductor shall be run to each service disconnecting means.

2. It must be bonded to each disconnecting means enclosure.

3. This conductor must be routed with the phase conductors.

4. It must not be smaller than the required grounding electrode conductor specified in Table 250-94.

5. For service-entrance phase conductors larger than 1100 kcmil copper or 1750 kcmil aluminum, the grounded conductor shall not be smaller than 12½ percent of the area of the largest service-entrance phase conductor.

6. Where the service-entrance phase conductors are installed in parallel, the size of the grounded service conductor must be based on the equivalent area of the ungrounded parallel service conductors.

Three exceptions to the rule for running the grounded service conductor to the service or sizing the conductor are provided.

Exception No. 1: The grounded conductor is not required to be larger than the largest ungrounded service-entrance phase conductor.

Exception No. 2: For high-impedance grounded neutral systems grounding connection requirements. See Section 250-27.

Exception No. 3: Where more than one service disconnecting means are located in an assembly listed for use as service equipment, only one grounded conductor is required to be run to the assembly. It must be bonded to the assembly enclosure.

This grounded service conductor provides the vital purpose of a low impedance path for fault current to return to the service. It must be installed to each service disconnecting means re-

**Grounded Conductor to Listed
Service Equipment Assembly**

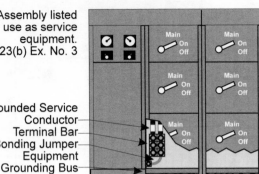

Assembly listed
for use as service
equipment.
250-23(b) Ex. No. 3

Grounded Service
Conductor
Terminal Bar
Main Bonding Jumper
Equipment
Grounding Bus

Figure 4-3

gardless of whether or not it is needed for the service.

Figure 4-4 illustrates a 3-phase power service taken from a grounded system. The grounded conductor is not needed for voltage requirements as only 3-phase or phase-to-phase connected loads are supplied. If the grounded conductor is not run to the service, as required in 250-23(b), we have a ground-fault circuit of high impedance as shown by the dashed lines. This is a violation of 250-51 and it becomes virtually impossible to clear a ground fault, thus introducing an unnecessary hazard in the system.

High and Low Impedance Paths

High Impedance Path

Low
Impedance
Path

High Impedance Path

Figure 4-4

However, if we comply with 250-23(b) and run the grounded conductor to the service, we have a low-impedance ground-fault path and have also complied with 250-51. In the latter case, the safety of the service is improved immeasurably. Since the grounded conductor is run only to provide a ground-fault path, the size of the conductor to be run will depend on the size of the phase conductors in the service. This conductor must not be smaller than the grounding electrode conductor given in Table 250-94.

Power and Lighting Service for 3-Phase System

Figure 4-5

Figure 4-5 represents the relationship between the service and equipment ground in a 3-phase, 4-wire delta system with the midpoint of one transformer grounded. Such systems are intended to supply economically both a 3-phase, 3-wire power service and a single-phase, 3-wire lighting service from one transformer bank. The lighting transformer must be big enough to supply all the lighting load plus the three-phase load. The other transformers are sized to carry the 3-phase, 3-wire load.

In the case of the 3-phase, 3-wire power service, there is no need for the neutral, as far as meeting voltage or phase requirements on the power service is concerned. However, it is required to run the grounded conductor of the system to the power service equipment and to use the main bonding jumper if we are to satisfactorily clear a ground fault developing on the power supply system.

Location of service grounding connection.

Section 250-23(a) requires that a grounding electrode conductor be used to connect the grounded service conductor to a grounding electrode. The connection to the grounded service conductor must be at an accessible location from the load end of the service drop or service lateral to, and including, the terminal bus at the service disconnecting means. These locations include current-transformer cans, meter enclosures, pull and junction boxes, busways, auxiliary gutters and wireways.

Service Grounding Connection to be Accessible

Figure 4-6

Many inspection authorities and serving utilities will not permit the grounding electrode connection to the system grounded conductor to be within current-transformer cans, meter enclosures or other enclosures that are sealed by the utility. Some utility policies prohibit connection within metering equipment, and many inspection authorities prohibit that connection location as they interpret the connection to be no longer "accessible" as utilities seal the metering equipment to prevent unauthorized access. It is important that serving utilities are consulted to be certain their system grounding policies are complied with.

The most practical and commonly accepted location for the grounding electrode conductor connection to the grounded service conductor is within the service disconnecting means or within a wireway at the service equipment location. The connection is usually made to the neutral bus.

Low-impedance grounding electrode conductor connections.

To obtain sufficiently low impedance, the point of connection of the grounding electrode conductor must be within the same metallic enclosure as the phase conductors and the grounded conductor.

The most practical point is within the service equipment.

The connection may be at one of the locations shown in Figure 4-6. Locations indicated are within the service panelboard, to the grounded conductor terminal block inside the current transformer cabinet, to the grounded service conductor terminal block inside the service wireway, or within the meter base.

It is also permissible to make the service grounding connections within each service disconnecting means rather than within the wireway. Means and methods of making and sizing these conductors are covered in Chapter Seven.

An additional grounding connection to an electrode must be made outside the building, usually at the transformer. This connection is usually provided by the service utility.

Section 250-23(a) of the Code goes on to disallow grounding of the grounded circuit conductor at any point beyond the service. Six exceptions are provided, several of which will be covered in later chapters of this text. The use of the words "grounded circuit conductors" is intended to cover all conductors whether they are feeders or circuits.

Sizing of grounded service conductor.

The size of the grounded service conductor (often the neutral) varies with the load as calculated in accordance with Section 220-22. This becomes the minimum size grounded service conductor unless modified by Section 250-23(b).

First of all, a load calculation in accordance with Section 220-22 should be performed. Then, the minimum size of conductor must be compared with Table 250-94. Section 250-23(b) requires the grounded conductor to be no smaller than the grounding electrode conductor as determined in Table 250-94. If the load calculation permits a conductor smaller than provided for in Table 250-94, the conductor must be increased in size to be not smaller than provided for in the table. If the load calculation requires a conductor larger than required by Table 250-94, the larger conductor must be used.

For example, if a 400 ampere service is to be installed from a grounded system, 500 kcmil copper conductors are selected for the ungrounded service conductors. By referring to Table 250-94, we see that the grounded service conductor can be no smaller than 1/0 copper or 3/0 aluminum. This conductor must be routed with the phase conductors and bonded to the service disconnecting means

Table 250-94**
Grounding Electrode Conductor for AC Systems

Size of Largest Service-Entrance Conductor or Equivalent Area for Parallel Conductors		Size of Grounding Electrode Conductor	
Copper	Aluminum or Copper-Clad Aluminum	Copper	*Aluminum or Copper-Clad Aluminum
2 or smaller	1/0 or smaller	8	6
1 or 1/0	2/0 or 3/0	6	4
2/0 or 3/0	4/0 or 250 kcmil	4	2
Over 3/0 thru 350 kcmil	Over 250 kcmil thru 500 kcmil	2	1/0
Over 350 kcmil thru 600 kcmil	Over 500 kcmil thru 900 kcmil	1/0	3/0
Over 600 kcmil thru 1100 kcmil	Over 900 kcmil thru 1750 kcmil	2/0	4/0
Over 1100 kcmil	Over 1750 kcmil	3/0	250 kcmil

Where multiple sets of service-entrance conductors are used as permitted in Section 230-40, Exception No. 2, the equivalent size of the largest service-entrance conductor shall be determined by the largest sum of the areas of the corresponding conductors of each set.

Where there are no service-entrance conductors, the grounding electrode conductor size shall be determined by the equivalent size of the largest service-entrance conductor required for the load to be served.

*See installation restrictions in Section 250-92(a).

(FPN): See Section 250-23(b) for size of alternating-current system grounded conductor brought to service equipment.

It is important to note that the minimum size of grounded service conductor that must be run to the service disconnecting means is based on the size of the ungrounded (hot) service-entrance conductors and not on the rating of the circuit breaker or fuse that is installed in the service. It is helpful to remember that Table 250-94 applies up to the service, and Table 250-95 applies beyond the service.

Dwelling unit services and feeders.

Special rules for sizing service entrance conductors and feeders are provided in Note Three to the Ampacity Tables of 0 to 2000 volts. These rules apply only to 120/240 volt, 3-wire, single-phase dwelling services and feeders. By following the conditions of this note, the specified size of service-entrance or feeder conductors shown in the table is permitted to be used based upon the service or feeder rating.

The grounded conductor (usually a neutral) is permitted to be smaller than the ungrounded (hot) conductors provided the rules of Sections 215-2, 220-22 and 230-42 are met.

Section 215-2 provides that the feeder neutral must be adequate for the load, must be a minimum size for certain loads and does not have to be larger than the service-entrance conductors that supply them.

Section 220-22 provides the method for calculating the feeder neutral load. The basic requirement is that the neutral must carry the maximum unbalanced load from the ungrounded conductors.

Section 230-42 generally requires that service-entrance conductors be of sufficient size to carry the loads as computed by Article 220. For grounded conductors in Section 230-42(c), the conductor cannot be smaller than required by Section 250-23(b).

The previous permission to size the grounded conductor not more than two sizes smaller than the ungrounded conductors given in the table in Note 3 has been removed from the Code.

enclosure. This is the minimum size grounded service conductor permitted and must be installed as noted even though there is no neutral load on the system or the calculated neutral load would permit a smaller conductor. Follow a similar procedure for other service installations.

Grounded service conductor based on size of ungrounded service conductors.

Figure 4-7

Parallel service conductors.

Where parallel service conductors as permitted by Section 310-4 are installed, the minimum size grounded service conductor is determined by multiplying the circular mil area of the conductors installed by the number of conductors installed in parallel.

Parallel Service Conductors

Parallel conductors must be:
1. Same length.
2. Same conductor material.
3. Same size in circular mil area.
4. Same insulation type.
5. Terminated in the same manner.

Figure 4-8

Example No. 1.

Given: Three, 4/0 copper conductors per phase are installed in parallel (electrically connected at each end to form a single conductor). Since we cannot multiply the size 4/0, we must convert the 4/0 conductors to their circular mil area.

- Refer to Chapter 9, Table 8 to determine the circular mil area of the 4/0 conductors. There we find the area to be 211,600 circular mils.
- Three times 211,600 = 634,800 circular mils.
- By referring to Table 250-94, we find the minimum size grounded service conductor is 2/0 copper or 4/0 aluminum.

This example assumes that all of the conductors are installed in the same raceway. This may not be practical due to the requirement that the derating or ampacity adjustment factors of Note 8 to the Ampacity Tables of 0 to 2000 Volts be applied. This obviously has the effect of requiring larger conductors to be installed than if the individual sets of service entrance conductors were installed in individual conduits or raceways.

Where installed in separate metal raceways, Section 300-3(b) requires that the neutral conductor must be installed in each raceway. In this case, Section 310-4 requires that the paralleled conductors be not smaller than 1/0, so the minimum size grounded conductor permitted in parallel is 1/0.

Example No. 2.

If we were to use six 4/0 conductors in parallel, the minimum size grounded service conductor is determined as follows:

We found the circular mil area of the 4/0 conductors to be 211,600 in Chapter 9, Table 8.

- Six times 211,600 = 1,269,600 circular mils. Since we have exceeded the 1100 kcmils given in Table 250-94, we must follow the rules in Section 250-23(b). There we find a requirement that the grounded service conductor be not smaller than 12½ percent of the equivalent area for parallel conductors.
- 1,269,600 times 12½ percent = 158,700 circular mils.
- By again referring to Chapter 9, Table 8, we find the next size larger than 158,700 circular mils to be a 3/0 conductor which has a circular mil area of 167,800.

This example also assumes that all the service entrance conductors are installed in the same raceway, which, due to derating requirements, may not be too practical.

Again, where installed in separate metal raceways, a grounded service conductor must be installed in each raceway and must not be smaller than 1/0 AWG. The grounded service conductor must also comply with Section 230-42(a). This section requires that the conductor be adequate to carry the load as determined by Article 220.

Section 220-22 requires that the neutral be sized for the maximum unbalance of the load. Examples of neutral conductor load calculations are found in Chapter 9 of the National Electrical Code®. In addition, where the length of run of the grounded conductor from the transformer to the service equipment is long, the size of grounded conductor should be increased.

As illustrated in Figure 4-9, for only underground installations in nonmetallic raceways, it is permitted by Section 300-5(i) Exception No. 2 to install all the conductors of each phase in the same raceway. To comply with this section, all the ungrounded conductors of phase A are installed in one raceway, phase B in another, C in the third and the grounded service conductors in another. This method is often chosen to allow phase conductors to more readily line up with bus terminations in switchboards. As can be seen in the illustration, this reduces the "rats nest" in the bottom of these enclosures caused by many conductors criss-crossing each other for termination. Another advantage is it is much easier to comply with the requirement that parallel conductors be the same length.

Care must be taken to eliminate the inductive

Underground Parallel Conductors

Service Equipment

Service Equipment

Metallic Raceways, All phase wires and neutral in same raceway.

Nonmetallic Raceways, All phase wires and neutral OK in individual raceways. 300-5(i) Ex 2

Figure 4-9

heating of metal enclosures with magnetic properties by cutting slots in the metal between the conduit entries or arranging for the manufacturer of the equipment to install a nonmetallic plate in the bottom of the equipment where the conduits terminate. [Section 300-20(a) and (b).] Another option is to terminate these conduits above the floor in the compartment of an open-bottom, floor-standing switchboard.

See Section 384-10 where the conduit or raceways, including their end fittings, are not permitted to rise more than 3 in. above the bottom of the enclosure.

Two or More Services to One Building from Grounded System

400 Ampere   100 Ampere

Size of grounded service conductor to each service based upon the size of ungrounded service conductor to the individual service.

Figure 4-10

Multiple services to one building.

Section 230-2(a) permits several services to one building under one of several exceptions to the basic rule. Each service that is supplied from a grounded system must be provided with a grounded service-entrance conductor. The size of ungrounded service-entrance conductor for each service determines the minimum size of grounded service conductor for that service. Each service is considered individually for sizing the grounded service conductor to it.

For example, a building has a 400 ampere and a 100 ampere service. The size of grounded service conductor is determined as follows:

400 ampere service.
- 750 kcmil THW aluminum ungrounded service conductors.
- Table 250-94 = 1/0 copper or 3/0 aluminum grounded service conductor.

100 ampere service.
- No. 2 copper ungrounded service conductors.
- Table 250-94 = No. 8 copper or No. 6 aluminum grounded service conductor.

Again, we must emphasize that this method determines the minimum size of grounded service conductor to comply with Section 250-23(b). A larger conductor may be required to carry the maximum unbalanced load on the neutral conductor as determined by Section 220-22.

High-impedance grounded systems.

Industrial plants having a continuous industrial process often have a need for uninterrupted electrical power and systems. It is quite common to see these plants located near a power company substation and to have more than one high voltage feeder to improve system reliability.

Another step that is commonly taken is to install high-impedance grounded systems rather than solidly grounded systems. Advantages include improved reliability, the ability to have ground-fault relaying and fault indicating alarms, as well as fewer problems to the system from transient overvoltages.

Four conditions must be met before the Code will permit high-resistance grounded systems to be installed. They are as follows:

High-Impedance Grounded System

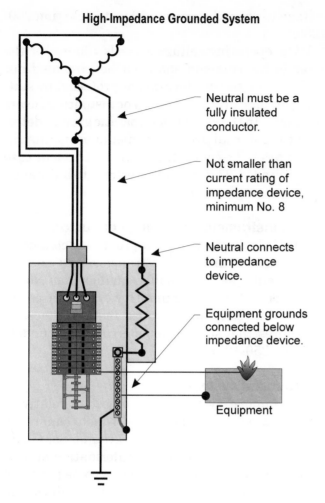

Neutral must be a fully insulated conductor.

Not smaller than current rating of impedance device, minimum No. 8

Neutral connects to impedance device.

Equipment grounds connected below impedance device.

Equipment

Figure 4-11

1. Qualified persons must be available to service and maintain the system.
2. Continuity of power is required.
3. Ground detectors must be installed to indicate an insulation failure.
4. Line-to-neutral loads are not served.

Specific rules are provided in Section 250-27 for installing these systems. The grounding impedance, usually a resistor, is installed between the transformer supplied grounded service-entrance conductor and the grounding electrode. Usually, the impedance device is sized to a value just greater than the capacitive charging current of the system.

A fully insulated grounded service-entrance conductor must be run to the service. This conductor must have an ampacity not less than the maximum current rating of the grounding impedance. The minimum size grounded service conductor cannot be smaller than No. 8 copper or No. 6 aluminum or copper-clad aluminum. Since the grounding impedance usually will limit the fault current to a fairly low value, the minimum size of neutral conductor is primarily related to physical protection, not to its current-carrying capabilities.

It is not required that the grounded service conductor be routed with the ungrounded service-entrance conductors since it will carry very little current in the event of a fault. [Section 250-27(d)] At the service, it is connected to the impedance device rather than to the service enclosure. An unspliced equipment bonding conductor is installed from the service enclosure where a terminal bar is installed for connection of equipment grounding conductors to the load side of the impedance device. [Section 250-27(e)] The grounding electrode conductor is permitted to be connected to any point from the grounded side of the grounding impedance to the equipment grounding connection at the service equipment or the first system disconnecting means. [Section 250-27(f)]

Second Ground Fault on High-Impedance Grounded System

High-impedance path through metal enclosures, loose raceway connections, etc.

Ground fault

Ground fault

Figure 4-12

This system is designed to limit the fault current on the first ground fault that might occur on the system. As can be seen in Figure 4-11, the impedance device is in series with the first ground fault. The electrical system will continue to function normally with the first ground fault present on the system. The ground-detection system will indicate the presence of the faulted condition by either a visual or audible signal or both. This is intended to alert maintenance personnel of the ground-fault condition so corrective action can be taken, hopefully during a period the plant is not planned to be in operation.

A second ground fault (illustrated in Figure 4-12) that occurs before the first fault is cleared will be a line-to-line or phase-to-phase fault that would be cleared by the service or feeder overcurrent device which will result in a power outage. This can, and at times has, involved two pieces of equipment in separate parts of the plant supplied by different feeders.

Instrument transformers, relays, etc.

Part L of Article 250 deals with grounding requirements for instrument transformers, relays, etc. The Code requires the grounding of secondary circuits of current and potential instrument transformers where the primary windings are connected to circuits of 300 volts or more to ground and, where on switchboards, shall be grounded irrespective of voltage. (Section 250-121) If a switchboard has no live parts or wiring exposed or accessible to other than qualified persons, such circuits need not be grounded where the primary windings are connected to circuits of 1000 volts or less.

Instrument transformer cases or frames are required to be grounded where accessible to other than qualified persons. Such cases or frames of current transformers need not be grounded where the primaries are not over 150 volts to ground and the current transformers are used exclusively to supply current to meters. (Section 250-122)

The cases of instruments, meters and relays, where operated with windings or working parts at 1000 volts or less, are required to be grounded. If such instrument cases and relays are not on switchboards and are accessible to other than qualified persons, the cases or other exposed metal parts shall be grounded if they operate at 300 volts or more to ground. [Section 250-123(a)]

For dead-front switchboards having no live parts on the front of the panel, such instrument cases shall be grounded. [Section 250-123(b)]

For live-front switchboards which have exposed live parts on the front of the panel, such instrument cases are not required to be grounded. However, the Code recommends the use of an insulating mat for the protection of the operator where the

voltage to ground exceeds 150 volts. [Section 250-123(c)]

If the operating voltage exceeds 1000 volts, the cases of instruments shall not be grounded but shall be isolated by elevation or protected by suitable barriers, grounded metal or insulating covers or guards. In the case of electrostatic ground detectors, the internal ground segments of the instruments are to be connected to the instrument case and grounded. The ground detector is to be isolated by elevation. (Section 250-124)

Size of instrument grounding conductor.

The Code specifies that the grounding conductor for circuits of instrument transformers and for instrument cases shall not be smaller than No. 12 copper or No. 10 aluminum. However, if cases are mounted directly on grounded metal surfaces of enclosures, they shall be considered to be grounded. (Section 250-125)

Ungrounded systems.

Ungrounded systems are subject to relatively severe transient overvoltages which can reach several times normal voltage to ground. See Chapter Two of this text for more information on this subject. Such abnormal voltages become potential hazards and often cause insulation failure and equipment breakdowns. If a system has one conductor grounded and is thus a grounded system, the value of such transient overvoltages as they develop is greatly reduced.

Grounding of Service for Ungrounded Systems

Connect service disconnect for ungrounded system to grounding electrode at building or structure.

Figure 4-13

An ungrounded system must have its conductor and equipment enclosures connected to a grounding electrode system at the building or structure served. This keeps such enclosures as near to ground potential as possible and reduces shock hazards to a minimum. These service equipment enclosures are grounded by connecting them to a grounding electrode system per Section 250-53. Grounding electrode systems are treated extensively in a later chapter of this text.

Hazard of operating a service ungrounded from a grounded system.

Figure 4-14 illustrates the hazard of operating an ungrounded service from a grounded system. The original ungrounded service on the right was installed some time before the service on the left. The first and original service was supplied by an ungrounded utility system. The service and feeder shown supplying equipment were protected by large overcurrent devices.

Hazard of Ungrounded Service from Grounded System

Additional Service

Original Service

Figure 4-14

Some time later, the service on the left which included a grounded service conductor was installed. At that time, the serving utility grounded the transformer bank and extended a grounded system conductor to the customer's service weatherhead. However, the grounded service conductor was not extended nor connected to the older, existing service. The older, existing service continued to supply power to a portion of the building, though without a grounded conductor connection.

The two services were connected with properly-sized grounding electrode conductors to a common grounding electrode system which included a metal water piping system.

Some time later a ground fault occurred in the equipment being supplied from the ungrounded service equipment. Since the system supplying the

equipment was grounded, current attempted to find its way back to the source of power to complete the circuit. Current flowed over the grounding electrode conductor at the ungrounded service, to the grounding electrode, through the grounding electrode to the grounded service and returned to the transformer bank over the grounded service conductor. In this and similar situations, the current will seek a path back to the system through all available paths and will divide based upon the impedance of the paths that are available.

A great deal of current flowed over the grounding electrode conductors that were not designed or installed for that amount of current. The resulting fire burned a great deal of the industrial plant to the ground.

The action necessary to prevent the fire was to run a properly-sized grounded system conductor to the existing service when the new service was installed. That conductor should have been extended to the service disconnecting means and bonded to it. That conductor would have provided a low-impedance path back to the transformer bank and allowed the overcurrent devices to clear the ground fault.

Methods of clearing ground faults and short circuits are discussed in detail in Chapter 11.

Chapter Four: The questions included here were developed using material included in this chapter. The answers can be found by reviewing the text. It is also important that students make use of the 1996 NEC®, where many answers can be found. See page 236 for anwers.

1. The grounded conductor is required to be run to each ac service disconnecting means where they operate at less than ____ volts and are grounded at any point.
   a. 2000
   b. 1500
   c. 1000
   d. 1200

2. Where an ac system operating at less than ____ volts is grounded at any point, the grounded conductor is required to be bonded to each disconnecting means enclosure.
   a. 1500
   b. 1200
   c. 2000
   d. 1000

3. Where an ac system operating at less than ____ volts is grounded at any point, the grounded conductor is required to be routed with the phase conductors.
   a. 1000
   b. 1200
   c. 1300
   d. 2000

4. Where an ac system operating at less than 1000 volts is grounded at any point the grounded conductor cannot be smaller than the required _____ conductor specified in Table 250-94.
   a. service-entrance
   b. grounding-electrode
   c. equipment grounding
   d. ungrounded service

5. For service-entrance phase conductors larger than 1100 kcmil copper or 1750 kcmil aluminum, the grounded conductor cannot be smaller than ____ percent of the area of the largest service-entrance phase conductor.
   a. 12½
   b. 14½
   c. 13¾
   d. 12

6. Where the service-entrance phase conductors are installed in ____, the size of the grounded service conductor is required to be based on the equivalent area of the ungrounded service conductors.
   a. a raceway
   b. a trench
   c. parallel
   d. a cable tray

7. Where more than one service disconnecting means is located in an assembly listed for use as service equipment, only one grounded conductor is required to be run to the assembly. It is required to ____ to the assembly enclosure.
   a. be sized at No. 8 and bonded
   b. not be bonded
   c. be bonded
   d. be sized at No. 6 and bonded

8. Some utility policies prohibit connection within metering equipment, and many inspection authorities prohibit that connection location because they interpret the connection to be no longer ____.
   a. accessible
   b. identified
   c. serviceable
   d. visible

9. The grounding electrode conductor connection to the grounded service conductor is most often made within the service disconnecting means and is connected to the ____ bus.
   a. unidentified
   b. isolated ground
   c. floating neutral
   d. neutral

10. Where the transformer supplying the service is located outside the building an additional grounding connection to an electrode is required to be made outside the building, usually at the ____.
   a. transformer
   b. pole
   c. meter
   d. grid

11. Where copper conductors sized at 500 kcmil are selected for the ungrounded service conductors supplying a 400 ampere service, and are installed from a grounded system, the size of the copper grounding electrode conductor cannot be less than No. ____.
   a. 6
   b. 8
   c. 1/0
   d. 4

12. Article 310, Note 3 to the Ampacity Tables of 0 to 2000 volts requires or permits the grounded service conductor for dwelling services and feeders to be:
   a. two sizes smaller than the ungrounded conductors.
   b. used for 60-ampere feeders.
   c. sized by calculation from Section 220-22.
   d. Type TW conductors.

13. Industrial plants having a continuous industrial process often have a need for ____ electrical power and systems.
   a. isolated
   b. larger
   c. five-wire, two-phase
   d. uninterrupted

14. Certain conditions must be met before the NEC® will permit high-resistance grounded systems to be installed. They include ____.
   a. qualified persons must be available to service and maintain the system.
   b. continuity of power is required, line-to-neutral loads are not served.
   c. ground detectors must be installed to indicate an insulation failure.
   d. all of the above.

15. Secondary circuits of current and potential instrument transformers where the primary windings are connected to circuits of ____ volts or more to ground and, where on switchboards, are required to be grounded irrespective of voltage.
   a. 300
   b. 100
   c. 200
   d. 150

16. Where a switchboard has no live parts or wiring exposed or accessible to other than qualified persons, such instrument transformer secondary circuits are not required to be grounded where the primary windings are connected to circuits of ____ volts or less.
   a. 480
   b. 1000
   c. 277
   d. 150

17. Instrument transformer cases or frames are required to be grounded where accessible to other than qualified persons. Such cases or frames of current transformers are not required to be grounded where the primaries are not over ____ volts to ground, and the current transformers are used exclusively to supply current to meters.
   a. 100
   b. 120
   c. 130
   d. 150

Chapter 5

Service and Main Bonding Jumpers

Objectives

After studying this chapter, the reader will be able to understand

- Definitions of bonding and bonding jumpers.
- Functions of the main bonding jumper.
- Sizing of main and equipment bonding jumpers.
- Methods for bonding at service equipment.
- Use of neutral for bonding on line side of service.
- Requirements for grounding and bonding of remote metering.

Definitions.

Bonding: "The permanent joining of metallic parts to form an electrically conductive path that will ensure electrical continuity and the capacity to conduct safely any current likely to be imposed."*

Bonding jumper, Main: "The connection between the grounded circuit conductor and the equipment grounding conductor at the service."*

Main bonding jumper.

The main bonding jumper is one of the most critical elements in the grounding system. This conductor is the link between the grounded service conductor, the equipment grounding conductor and the grounding electrode conductor. This main bonding jumper is called upon to carry all the fault current from the service enclosure as well as from the equipment grounding system that is returning to the source. Section 250-53(b) requires that an unspliced main bonding jumper be installed at each service disconnect.

Main Bonding Jumper

Neutral bus

Main bonding jumper may be wire, bus or screw

Equipment grounding bus bonded to enclosure

Figure 5-1

The main bonding jumper is permitted to consist of a wire, bus, screw or other suitable conductor. It must be fabricated of copper or other corrosion-resistant material. Where the main bonding jumper consists of a screw, it must have a green finish that is visible with the screw installed. This will assist in identifying the bonding-jumper screw from other screws that are in or near the neutral bus.

Where more than one service disconnecting means is located in an assembly listed for use as service equipment, Section 250-53(b) Exception No. 1 permits one grounded service conductor to be run to the enclosure and be bonded to the assembly enclosure. This means that only one main bonding connection is required to be made to the assembly. The sections of the assembly are bonded together by means of an equipment grounding conductor bus or by being bolted together. This equipment usually consists of listed switchboards, panelboards or motor control centers.

Exception No. 2 to Section 250-53(b) permits the bonding for high-impedance grounded neutral systems to be performed differently. See Chapter Four of this text for methods and requirements for grounding high-impedance grounded neutral systems. Also see NEC® Section 250-27.

Functions of Main Bonding Jumper.

The main bonding jumper performs three major functions. Those functions include:

1. Connecting the grounded service conductor to the equipment grounding bus or conductor.
2. Providing the low-impedance path for the return of ground-fault currents to the grounded service conductor.

Major Functions of
Main Bonding Jumper

Grounded Service Conductor

Neutral bus
Main bonding jumper

Equipment Grounding Bus

Figure 5-2

3. Under certain conditions given in Section 250-23(a) Exception No. 5, it is permitted to connect the grounding electrode conductor to the equipment grounding terminal bar. In this case, the main bonding jumper serves to connect the grounded service conductor to the grounding electrode.

This main bonding jumper completes the ground-fault return circuit from the equipment through the service to the source as is illustrated in Figure 5-2.

Size of main bonding jumper in listed enclosures.

Where a listed switchboard or panelboard is installed, the main bonding jumper that is provided with the equipment is rated for the size of conductors that would be used for the service. The proper size of the main bonding jumper is verified by the listing agency and can be used without calculating the size. For example, if a bus bar, strap, conductor or screw is furnished by the manufacturer as part of the listed equipment, it may be used without calculating its adequacy. Manufacturers are required to provide the main bonding jumper by Section 384-3(c). The method for sizing of the main bonding jumper in listed service equipment is found in Underwriters Laboratories Safety Standard for the equipment under consideration.

Size of main bonding jumper at single service-disconnect or enclosure.

Since the main bonding jumper must carry the full ground-fault current of the system back to the grounded service conductor (which may be a neutral), its size must relate to the rating of the service conductors which supply the service. The mini-

mum size of the main bonding jumper is found in Table 250-94 as required by Section 250-79(d).

For example, where 250 kcmil aluminum service-entrance conductors are installed, the main bonding jumper is found to be No. 4 copper or No. 2 aluminum by reference to Table 250-94.

The size of the main bonding jumper does not directly relate to the rating of the service overcurrent device. Do not attempt to use Table 250-95 for this purpose. Table 250-95 gives the minimum size of equipment grounding conductors for feeders and circuits on the load side of the service.

Sizing main bonding jumper for parallel service conductors.

Where parallel conductors are installed, Section 250-79 requires that the circular mil area of the conductors are added together to determine the minimum size main bonding jumper required. For example, where four 250 kcmil conductors are installed in parallel, they are treated as a single 1000 kcmil conductor. By reference to Table 250-94 we find the main bonding jumper, if aluminum service-entrance conductors are used, would be 2/0 copper or 4/0 aluminum.

Main Bonding Jumper for Parallel Runs

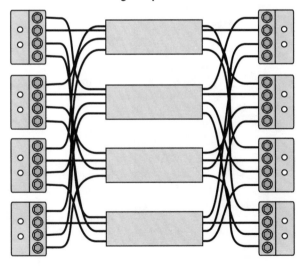

Four 250 kcmil aluminum conductors per set
4 x 250 = 1000 kcmil
Refer to Table 250-94
Select 2/0 copper or 4/0 aluminum

Figure 5-4

Where the service-entrance conductors are larger than the maximum given in Table 250-94, Section 250-79 requires the main bonding jumper to be not

Main Bonding Jumper at Single Disconnect

Grounded Electrical System

Service-entrance conductors

Size of main bonding jumper based on size of ungrounded service-entrance conductor

Figure 5-3

less than 12½ percent of the area of the largest phase conductor.

This is illustrated by the following example: Three 500 kcmil copper conductors are installed as service-entrance conductors.

3 x 500 kcmil = 1500 kcmil.

1500 x .125 = 187,500 circular mils.

Since a 187,500 circular mil conductor is not a standard size, we next refer to Chapter 9, Table 8 to find the area of conductors.

The next conductor exceeding 187,500 circular mils is a 4/0 AWG conductor which has an area of 211,600 circular mils.

Follow a similar procedure for determining the minimum size main bonding jumper required for other sizes of parallel service-entrance conductors.

Sizing of equipment bonding jumper on line side of service.

Equipment bonding jumpers on the line side of the service and main bonding jumper must be sized to comply with Table 250-94. This is required by Section 250-79(d). For example, where 250 kcmil copper conductors are installed as service-entrance conductors, Table 250-94 requires a No. 2 copper or 1/0 aluminum bonding jumper.

Where conductors are used in parallel and the sum of the circular mil area exceeds 1100 kcmil copper or 1750 kcmil aluminum, the equipment bonding conductor must be not less than 12½ percent of the area of the ungrounded phase conductors.

Size of Equipment Bonding Jumper on Line Side of Service

Neutral Bus

Where installed in series, equipment bonding jumper sized for total circular mil area of service-entrance conductors in parallel.

Open bottom switchboard

Figure 5-5

For example, if five 250 kcmil copper conductors in parallel are run to a service, the equipment bonding jumper for bonding metal raceways must not be smaller than 3/0 copper.

This is determined as follows:

Five x 250 kcmil = 1250 kcmil.

1250 kcmil x .125 = 156,250 circular mils.

The next larger conductor found in Chapter 9, Table 8 is 3/0 with an area of 167,800 circular mils.

In this case, a 3/0 copper bonding conductor must be connected to each metal raceway in series between each metal raceway and the grounded service enclosure or conductor.

Individual Bonding Jumpers

Service equipment enclosure

Neutral or Equipment Grounding Bus

Size bonding jumpers per conductors in raceway

Figure 5-6

A more practical method of performing the bonding for multiple raceway services may be to connect an individual bonding conductor between each raceway and the enclosure or grounded service conductor. This is permitted by Section 250-79(d). As can be seen, this will usually result in a smaller conductor which is easier to install.

Based upon Table 250-94, the minimum size bonding conductor for the individual raceways containing 250 kcmil copper service-entrance conductors is No. 2 copper or 1/0 aluminum.

Different conductor material.

Section 250-79(d) provides instructions on sizing the main bonding jumper or equipment bonding jumper on the supply side of the main bonding jumper where different conductor materials are used for the service-entrance conductors and the bonding jumper. The rule is that phase conductors of the same material (copper or aluminum) as the bonding jumper, having equivalent ampacity to those that are installed, are assumed. This is illustrated as follows:

Assume aluminum phase conductors and a copper bonding jumper.

Three 750 kcmil Type THW aluminum conductors are installed.

From Table 310-16, 385 amperes x 3 = 1155 amperes. The smallest type THW copper conductor that has an equivalent rating is 600 kcmil with an ampacity of 420.

Next, determine the total circular mil area of the copper conductors.

Three x 600 kcmil = 1800 kcmil.

1800 kcmil x .125 = 225 kcmil.

The next standard size is 250 kcmil copper which is the minimum size bonding jumper permitted to bond equipment at or ahead of the service equipment in this example.

Bonding service equipment enclosures.

The Code requires that electrical continuity of service equipment and enclosures that contain service conductors be established and maintained by bonding. The items required to be bonded together are stated as follows in Section 250-71(a):

Bonding Service Equipment Enclosures

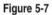

Raceways and cable armor or sheath

Service equipment enclosures including meter fittings

Figure 5-7

(a) The service raceways, cable-trays, cablebus framework or service cable armor or sheath.

(b) All service equipment enclosures containing service conductors, including meter fittings, boxes or the like, interposed in the service raceway or armor.

(c) Any metallic raceway or armor which encloses the grounding electrode conductor. (This subject is covered in detail in Chapter 7 of this text.)

An exception to this requirement for bonding at service equipment is found in Section 250-55. This section refers to an underground service cable which is metallically connected to the underground service conduit. The Code points out that if a service cable contains a metal armor, and if the service cable also contains an uninsulated grounded service conductor which is in continuous electrical contact with its metallic armor, then the metal

covering of the cable is considered to be adequately grounded.

Use of neutral for bonding on line side of service.

Section 250-61(a) permits the use of the grounded circuit conductor (may be the neutral) for grounding and bonding equipment on the line side of the service disconnecting means. (Two other applications of this bonding are explored in later chapters of this text.) Often, connecting the grounded service conductor to equipment such as meter bases, current transformer enclosures, wireways and auxiliary gutters is the most practical method of bonding these enclosures.

Use of Neutral for Bonding on Line Side of Service

Figure 5-8

Usually, self-contained meter sockets are produced with the grounded conductor terminals bonded directly to the enclosure. The meter enclosure is bonded by the connection of the grounded circuit conductor to these terminals. No further bonding conductor connection to the meter enclosure is required. Current from a ground fault to the meter enclosure will return to the source by the neutral and, hopefully, will allow enough current to flow in the circuit to operate the overcurrent protection on the line side of the utility or other transformer.

In addition, meter enclosures installed on the load side of the service disconnecting means are permitted by Section 250-61(b) to be grounded (bonded) to the grounded service conductor provided that:

a. service ground-fault protection is not installed; and

b. the meter enclosures are located near the

service disconnecting means. (No distance is used to clarify what is meant by the word "near.")

Means of bonding at service equipment.

The methods for bonding at service equipment are outlined in Section 250-72. These requirements for bonding are more restrictive at services than downstream from the service. The reason this is so important is service equipment and enclosures would be called upon to carry heavy fault currents in the event of a line-to-ground fault. The service conductors in these enclosures have only short-circuit protection provided by the overcurrent device on the line side of the transformer. Overload protection only is provided at the load end of the service conductor by the overcurrent device. That is one of the reasons the Code limits the length of service conductors permitted inside of a building.

Bonding of these enclosures is to be done by one or more of the following methods from Section 250-72:

(a) Bonding to the grounded service conductor through the use of listed lugs, pressure connectors, or other approved means which do not depend upon solder. (See Figure 5-8.)

Methods of Bonding Service Equipment

(b) Threaded couplings or bosses on enclosures made up wrenchtight.

Conduit hub furnished in many trade conduit sizes as accessory by equipment manufacturer. Install according to manufacturer's instructions.

Weatherproof conduit hub for field application. Cross-sectional view showing "O" ring seal and bonding screw. Manufacturer identifies it as being suitable for grounding and bonding.

(c) Threadless couplings and connectors made up tight for rigid metal conduit, intermediate metal conduit and electrical metallic tubing.

Figure 5-9

(b) Threaded couplings and threaded bosses in a rigid or intermediate metal conduit system where the joints shall be made up wrench tight.

(c) Threadless couplings and connectors are permitted where they are made up tight for rigid and intermediate metal conduit and electrical metallic tubing.

Note that the Code here states that "Standard locknuts and bushings shall not be used for the bonding required by this section." This does not intend to prevent the use of "standard" locknuts and bushings, it is just that they cannot be relied upon as the sole means of bonding that is required by this section. "Standard" locknuts are commonly used outside the enclosure on conduit to be bonded with a bonding bushing inside the enclosure.

Bonding Fittings

Bonding Bushing

Bonding Locknut

Bonding Wedge

Figure 5-10

(d) Bonding jumpers are required to be used around concentric or eccentric knockouts that are punched or otherwise formed so as to impair the electrical connection to ground.

(e) Other approved devices such as grounding-type locknuts.

Parallel bonding conductors.

Section 250-79(d) requires that the equipment bonding jumper be run in parallel where service-entrance conductors are paralleled in two or more raceways or cables and the equipment bonding jumper is routed with the raceways or cables.

In this case again, the size of the bonding jumper for each raceway is based upon the size of the service-entrance conductor in the raceway by referring to Table 250-94.

Parallel Bonding Conductors

Equipment bonding jumper must be installed in parallel where routed with phase conductors that are installed in parallel

Figure 5-11

Grounding and bonding of remote metering.

As mentioned before, Section 250-71(a) requires all equipment containing service conductors to be bonded together and to the grounded service conductor. This includes remote (from the service equipment) meter cabinets and meter sockets.

Grounding and Bonding of Remote Metering

Bond remote CT enclosure to neutral.
Supplemental grounding electrode permitted.

Service equipment

Figure 5-12

Grounding and bonding of equipment such as meters, current transformer cabinets and raceways to the grounded service conductor at locations on the line side of and remote from the service disconnecting means increases safety.

These equipments should never be grounded only to a grounding electrode such as a ground rod. Figures 5-12 and 5-13 show why. If a ground fault

occurred at this line-side equipment, the only means for clearing a ground fault would be through the grounding electrodes and earth. Given the relatively high impedance of this path through the earth and resistance of ground rods, little current will flow in this path. This leaves the equipment at a dangerous voltage above ground potential just waiting to shock or electrocute a person or animal that may contact them. The voltage drop across this portion of the circuit can easily be calculated by using Ohms Law. (Resistance times the current gives the voltage.) There are many records of livestock being electrocuted while contacting this equipment that was improperly grounded. Sections 250-51 and 250-91(c) require that the earth not be used as the sole equipment grounding conductor.

Bonding Remote Meter Loop

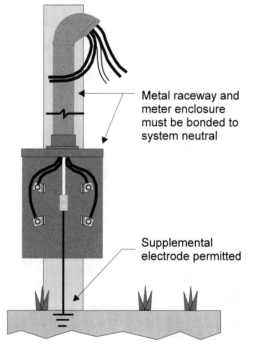

Metal raceway and meter enclosure must be bonded to system neutral

Supplemental electrode permitted

Figure 5-13

The most practical method for grounding and bonding the line-side equipment is to bond the grounded service conductor to it. As can also be seen in Figure 5-12 and 5-13, a ground fault to the equipment will have a low impedance path back to the source through the grounded service conductor. This will allow a large current to flow in the circuit to cause the overcurrent protection on the line side of the transformer to clear the fault.

Supplemental grounding electrode.

In accordance with Section 250-91(c), it is permissible to install a grounding electrode at the remote meter location shown in Figures 5-12 and 5-13 to supplement the grounded service conductor. The grounding electrode will attempt to keep the equipment at the earth potential that exists at the meter location. In addition, the electrodes at the remote meter and at the service are bonded together by the grounded service conductor installed between the metering and service equipment. This complies with Section 250-54 which requires a common grounding electrode or where two or more electrodes are installed, they be bonded together.

As previously stated, these grounding electrodes should never be used in an attempt to be the only means for grounding or bonding these enclosures. Some electric utilities require the grounding electrode installation and connection at these remote-metering installations.

More extensive discussion of this subject is found in Chapter Six of this text.

Bonding of multiple service disconnecting means.

Installation of multiple services as permitted by Section 230-2(a) and multiple disconnecting means can take several forms. Included are services for:

1. Fire pumps, emergency, legally required, standby, optional standby or parallel power production systems.
2. Two to six sets of service-entrance conductors, each serving individual service disconnecting means. See Section 230-40, Exception No. 2.
3. Two to six sets of service laterals running to the same location and connected together at their line end but not at their load end. See Section 230-2, Exception No. 7.
4. Buildings of more than one occupancy that have a set of service-entrance conductors run to each occupancy. See Section 230-40, Exception No. 1.
5. Where other services are required and as permitted by one or more of the other exceptions to Section 230-2(a).

The basic rule for sizing of the equipment bonding jumper for bonding these various configurations is found in Section 250-79(d). This section requires that the bonding jumpers on the line side of the main bonding jumper be sized per Table 250-94. Also, the size of the bonding jumper for each raceway is based on the size of service-entrance conductors in each raceway. As discussed earlier,

conductors larger than given in Table 250-94 are required for larger systems. Since different sizes of service-entrance conductors are installed at various locations, we size the equipment bonding conductor for the size of conductors at each location.

**Bonding of Multiple
Service Disconnecting Means**

Figure 5-14

For example, the appropriate size of bonding jumper for the installation in Figure 5-14 with the assumed size of conductors is as follows: (all sizes copper)

Service-Entrance Conductor	Bonding Jumper
a. 500 kcmil in service mast	1/0
b. 1000 kcmil in wireway	2/0
c. 300 kcmil to 300 ampere service	No. 2
d. 3/0 to 200 ampere service	No. 4
e. No. 2 to 125 ampere service	No. 8

A practical method for bonding the wireway (sometimes referred to as a "hot" gutter) is to connect the grounded service conductor directly to the wireway. This is done by bolting a multi-barrel lug directly to the wireway and connecting the neutral conductors to the lug. This connection is permitted by Section 230-46, Exception No. 2.

To comply with the rules in Sections 250-23(b) and 250-53(b), the grounded service conductor must also be extended to each service disconnecting means and be bonded to the enclosure.

Chapter Five: The questions included here were developed using material included in this chapter. The answers can be found by reviewing the text. It is also important that students make use of the 1996 NEC®, where many answers may be found. See page 236 for answers.

1. The permanent joining of metallic parts to form an electrically conductive path that will assure electrical continuity and the capacity to conduct safely any current likely to be imposed is defined as ____.
   a. grounding
   b. bonding
   c. welded
   d. grounded

2. The connection between the grounded circuit conductor and the equipment grounding conductor at the service is defined as the ____.
   a. main bonding jumper
   b. gounding electrode conductor
   c. equipment bonding jumper
   d. neutral conductor

3. The main bonding jumper is permitted to consist of a ____ or other suitable conductor.
   a. wire
   b. bus
   c. screw
   d. all of the above

4. Where the main bonding jumper consists of a ____ only, it is required to have a green finish that is visible with the ____ installed.
   a. bus
   b. screw
   c. wire
   d. jumper

5. Where ____ kcmil aluminum service-entrance conductors are installed, the main bonding jumper is required to be No. 4 copper or No. 2 aluminum.
   a. 1/0
   b. 2/0
   c. 3/0
   d. 250

6. Where the service-entrance conductors are larger than the maximum sizes given in Table 250-94, the main bonding jumper cannot be less than ____ percent of the area of the largest phase conductor.
   a. 9½
   b. 10½
   c. 11½
   d. 12½

7. Where conductors are used in parallel, and the sum of the circular mil area exceeds ____ kcmil copper or ____ kcmil aluminum, the equipment bonding conductor must be not less than 12½ percent of the area of the ungrounded phase conductors.
   a. 1100 - 1750
   b. 1000 - 1650
   c. 1050 - 1500
   d. 1075 - 1400

8. The Code requires that electrical continuity of the grounding circuit of service equipment and enclosures that contain service conductors be obtained by ____.
   a. grounding
   b. welding
   c. approval
   d. bonding

9. The grounded circuit conductor (may be neutral) is permitted to be used for grounding and bonding on the ____ side of the service disconnecting means.
   a. load
   b. line
   c. subpanel
   d. control center

10. Grounding and bonding of equipment such as meters, current transformer cabinets and raceways to the grounded service conductor at locations on the line side of and remote from the service disconnecting means increases ____.
   a. voltage
   b. current
   c. safety
   d. cost

Notes:

Chapter

6

Grounding Electrodes

Objectives

After studying this chapter, the reader will be able to understand

- Definition and general requirements for grounding electrodes.
- Grounding electrode system to be used.
- Sizing bonding jumper for grounding electrode system.
- Description and installation of made electrodes.
- Common grounding electrode.
- Objectionable current flow and resistance of grounding electrodes.

Definition.

Grounding electrode. A conductor used to establish a ground and to connect electrical systems and/or equipment to the earth.

Functions of Grounding Electrode

1. Connects the electrical system to earth.

2. Connects electrical equipment to earth.
3. Attempts to maintain equipment at earth potential.
4. Little effect in clearing ground-faults.

Figure 6-1

General.

Grounding electrodes provide the essential function of connecting the electrical system to the earth. The earth is considered to be at zero potential. In some cases, the grounding electrode serves to ground the electrical system. In other instances, the electrode is used to connect noncurrent carrying metallic portions of electrical equipment to the earth. In both situations, the grounding electrode attempts to maintain the electrical equipment at the earth potential present at the grounding electrode.

In Section 250-53(a), we find a requirement to connect the equipment grounding conductors, the service-equipment enclosures, and where the system is grounded, the grounded service conductor to a grounding electrode. The conductor used to make this connection is the grounding electrode conductor.

Grounding electrode system.

The Code in Section 250-81 requires that, where available on the premises at each building or structure served, all grounding electrodes including "made" electrodes be bonded together to form the grounding electrode system. This includes metal underground water pipe, metal frames of buildings, concrete-encased electrodes, and ground rings. A bonding jumper must be installed between the grounding electrodes to bond them together.

Grounding Electrode System

(a) Metal underground water pipe in direct contact with the earth for 10 ft. or more.

(b) Metal frame of the building where it is effectively grounded.

(c) Concrete encased electrode.

(d) Ground ring.

Figure 6-2

Where the interior metal water pipe is used as a part of the grounding electrode system or as a conductor to bond other electrodes together to create the "grounding electrode system," Section 250-81 requires that all bonding take place within the first 5 feet from the point the water pipe enters the building. This section does not require that the interior water pipe be used for the purpose of interconnecting other electrodes to form the grounding electrode system. Any of the other electrodes, such as the metal frame of the building, concrete encased electrode or ground ring, can be used for the purpose of interconnecting the other grounding electrodes. Where these other electrodes are used for this purpose, no restrictions are placed on where the connections are permitted to be made or how far inside the building they are permitted. Section 250-112 requires grounding electrode conductor connections to grounding electrodes to be accessible except for connections to a buried, driven or concrete encased electrode.

Grounding electrodes required to be used.

Several grounding electrodes are required to be used where "available on the premises at each building or structure served." The Code does not define what is meant by "available" nor does it require that the electrodes be made available where they are not. For example, if the building has encased concrete reinforcing rods when the electrical system is installed, it is not required that the rods be exposed for connection. On the other hand, the concrete reinforcing rods must always be used when "available." The grounding electrodes are not listed in a order of preference nor is it optional to choose which ones to use.

Electrodes that must be used, in addition to any "made" electrodes that exist at the building or structure served, where "available" are as follows:

Grounding Electrodes That Must Be Used

(a) Metal underground water pipe in direct contact with the earth for 10 ft. or more including any metal well casing that is effectively bonded to the pipe.

(b) Metal frame of the building where it is effectively grounded.

(c) Concrete encased electrode. Encased by at least 2 in. of concrete, 20 ft. or more of ½ in. or larger electrically conductive steel reinforcing rods or bars or at least 20 ft. of bare No. 4 copper conductor.

(d) Ground ring encircling building in direct contact with the earth. No smaller than No. 2 bare copper at least 20 ft. long and 2½ ft. deep.

Figure 6-3

1. Metal Underground Water Pipe. Defined in Section 250-81(a) as "A metal underground water pipe that is in direct contact with the earth for 10 feet or more including any metal well casing that is effectively bonded to the pipe." There is no minimum or maximum pipe size given. Types of metal, such as steel, iron, cast iron, stainless steel or even aluminum are not distinguished. Different types of water pipes such as for potable water, fire protection sprinkler systems, irrigation piping, etc., are also not defined. As a result, all of these metal underground water pipes must be used where "available at each building or structure served."

Continuity of the grounding path of the water pipe grounding electrode or the bonding of interior piping systems cannot depend on water meters or on filtering devices or similar equipment. Where a water meter or filtering equipment is in this metal water piping system, it must be bonded around to maintain continuity even if the water meter or filter is removed.

2. Metal Frame of the Building. Section 250-81(b) requires the metal frame of the building be used as a grounding electrode where it is effectively grounded. "Grounded effectively" is defined in Article 100 and means that the metal frame of the building is, "Intentionally connected to earth through a ground connection or connections of sufficiently low impedance and having sufficient current-carrying capacity to prevent the buildup of voltages that may result in undue hazards to connected equipment or to persons."

To be an effective grounding electrode, the metal frame of the building must have a sufficiently low-impedance contact with the earth to pass current when called upon to do so and to maintain the electrical system at or near the electrical potential of the surrounding earth. The building steel can be connected to the earth by bolted or welded connection to reinforcing steel in foundations or footings that are in turn encased in concrete. Also, building structural steel may itself be encased in concrete that is in contact with the earth.

The structural steel should be tested with an earth resistance tester if in doubt about its resistance to ground and adequacy as a grounding electrode.

3. Concrete-Encased Electrodes. Section 250-81(c) defines this grounding electrode as one or more steel reinforcing bars or rods that are not less than 20 feet in length and ½ inch in diameter or 20 feet or more of bare copper conductor not smaller than No. 4. These electrodes must be located within or near the bottom of the foundation or footing and be encased by at least 2 inches of concrete. These reinforcing rods are often welded together and form an excellent low resistance grounding electrode.

Reinforcing rods must be of bare, zinc galvanized or other electrically conductive coated steel material. Obviously, insulated reinforcement rods would not perform properly as a grounding electrode. Some complaints have been made that lightning surges, that are dissipated through this electrode, break out chunks of concrete where the surge exits the footing.

This grounding electrode is commonly referred to as the "Ufer ground" after H.G. Ufer who spent many years documenting its effectiveness. See the Appendix of this text for additional information on the development of the concrete-encased electrode.

Several electrical inspection agencies require that a concrete-encased electrode be installed or connected to the service prior to authorizing electrical service due to its effectiveness in most any climatic and soil condition.

4. Ground Ring. Section 250-81(d) recognizes a copper conductor, not smaller than No. 2 and at least 20 feet long, as a ground ring grounding electrode. The conductor must "encircle" the building or structure and be buried not less than 2½ feet deep. Ground rings often are installed at telecommunication central offices. Where available on the premises served, ground rings must be used as one or more of the grounding electrodes making up the grounding electrode system.

Supplemental electrode.

Section 250-81(a) requires that where the only grounding electrode available and connected at the building or structure served is a metal underground water pipe, it be supplemented by another grounding electrode. Electrodes suitable to supplement the metal underground water pipe are a concrete-encased electrode, ground ring, other local metal underground systems or structures, rod and pipe electrodes, or plate electrodes. Often, metal underground water pipes are replaced by plastic water services or the system continuity is interrupted by nonmetallic couplings or repairs.

The words "or the interior metal water piping at any convenient point" have been deleted from Section 250-81(a). As a result, the supplemental electrode, required where an underground metal water pipe is the only grounding electrode, will be permitted to be connected to only the grounding electrode conductor, the grounded service-entrance conductor, the grounded service raceway or to any grounded service enclosure. An exception to this section permits the bonding connection to the interior metal water piping in a qualifying industrial or commercial plant to be made at any location if the entire length of interior metal water pipe that is being used as a conductor is exposed.

Often, changes, repairs or modifications are made to the metallic water piping systems with nonmetallic pipe or fittings. In this case, it is possible to inadvertently isolate portions of the

Supplemental Grounding Electrode

Supplemental grounding electrode permitted to be connected to: Grounded service-entrance conductor, grounded metal service raceway, grounded service enclosure or, grounding electrode conductor. Supplemental grounding electrode

Figure 6-4

grounding system from the grounding electrode conductor. This is another in several steps that has been taken over recent years to reduce the emphasis and reliance on the metal water piping system for grounding of electrical systems.

Size of bonding jumper for grounding electrode system.

The bonding jumper used to bond the grounding electrodes together to form the grounding electrode system must be sized in accordance with Section 250-94 based on the size of the ungrounded service-entrance conductor.

Size of Bonding Jumper for Grounding Electrode System

Size 250-94   Max. No. 4

Max. No. 2

Figure 6-5

For example, if the service-entrance conductor is 500 kcmil copper, the minimum size of bonding jumper is determined by reference to Section 250-94, including the exception as follows:

- to metal underground water pipe and metal

frame of a building; No. 2 copper or No. 1/0 aluminum conductor.

- to "made" electrodes as in Section 250-83(c) or (d) such as pipes, rods or plates; that portion that is the sole connection to the made electrode; No. 6 copper or No. 4 aluminum [Exception (a)].
- to a concrete encased electrode as in Section 250-81(c); that portion that is the sole connection to the concrete encased electrode; No. 4 copper conductor [Exception (b)].
- to a ground ring as in Section 250-81(d); that portion that is the sole connection to the ground ring; - not larger than the ground ring conductor [Exception (c)].

Note that aluminum is not permitted to be installed as a grounding electrode conductor where in direct contact with masonry or the earth or where subject to corrosive conditions. Where used outside, aluminum or copper-clad aluminum grounding conductors are not permitted within 18 inches of the earth. [Section 250-92(a)]

No sequence for installing the bonding jumpers is given. However, the minimum wire size required to the various grounding electrodes must be observed. In addition, the unspliced grounding electrode conductor is permitted to run from the service equipment to any convenient grounding electrode. All the grounding electrodes are then bonded together to form the grounding electrode system.

Individual grounding electrode conductors are permitted to be installed from the service equipment to a grounding electrode rather than being installed in a circular manner. The minimum size of each grounding electrode conductor to the individual grounding electrode is shown in Figure 6-5.

Size of Individual Grounding Electrode Conductor

Figure 6-6

Section 250-112 requires that these bonding connections be accessible and made in a manner that will assure a permanent and effective ground. An exception from the requirement that the connections be accessible is provided for concrete-encased, buried or driven electrodes.

Made electrodes.

Where the electrodes described in Section 250-81 are not available at the service location, a grounding electrode must be "made." The made electrode as provided for in Section 250-83 may be local metal underground systems or structures, driven pipes, or rods or buried plates conforming to the following requirements:

Made Electrodes

Local metallic underground systems such as piping and tanks.

Pipe or conduit electrodes not less than 8 feet long. Min. ¾ inch trade size. Galvanized if of iron or steel. Rods of iron or steel at least ⅝ inch. Stainless steel, nonferrous rods or equal to be listed and not less than ½ inch.

Plate electrodes to expose not less than 2 sq. ft. of surfacre to exterior soil. Minimum ¼-inch thick if of iron or steel, .06 inch if nonferrous metal.

Figure 6-7

(1) Local systems. Local metallic underground systems as piping, tanks, etc.

(2) Pipe electrodes. Pipe or conduit electrodes shall be not less than 8 feet in length nor smaller than ¾-inch trade size and if of iron or steel, shall be galvanized or metal-coated for corrosion protection.

(3) Rod electrodes. Electrodes of steel or iron shall be at least ⅝-inch diameter. Rods of nonferrous metal or stainless steel that are less than ⅝-

inch in diameter shall be listed and be at least ½-inch in diameter.

(4) Plate electrodes. Electrodes shall have at least 2 square feet of surface in contact with exterior soil. If of iron or steel, the plate shall be at least ¼-inch thick. If of nonferrous metal, they shall be at least 0.06 inch thick.

Note that underground metal gas piping systems are not permitted to be used as a grounding electrode. This does not eliminate the requirement that interior metal gas piping systems be bonded. See Chapter 8 of this text for additional information on bonding of metal piping systems.

Installation of made electrodes.

Where practicable, made electrodes must be installed below permanent moisture level. This is a key ingredient in establishing an effective made electrode. They also are required to be free from nonconductive coatings such as paint and enamel.

Rod and pipe electrodes must be installed so at least 8 feet is in contact with the soil. They must be driven vertically unless rock bottom is encountered. If rock bottom is encountered which prevents the rod from being driven 8 feet vertically, the rod is permitted to be installed at an oblique angle of not more than 45 degrees from vertical or it can be buried in a trench that is at least 2½ feet deep.

Installation of Rod-Type Electrodes

Surface of the earth

8 feet

45° Max.

2-½ ft. Min.

Rock bottom

Figure 6-8

The upper end of the rod must be flush with or below ground level unless the above ground end of the rod and the grounding electrode attachment are protected from physical damage. This, of course, requires that a ground rod longer than 8 feet be used if any of the rod is exposed above ground level.

Section 250-117 requires that ground clamps or other fittings be approved (acceptable to the authority having jurisdiction) for general use without pro-tection or be protected from physical damage by metal, wood or equivalent protective covering.

Common grounding electrode.

Section 250-54 requires that a common grounding electrode be used for all system grounding in or at a building. In addition, where more than one service supplies a building, the common grounding electrode must be used for all services. This section recognizes that where two or more grounding electrodes are bonded together, they are considered to be one electrode.

Common Grounding Electrode

Effectively grounded building steel or interior metal water pipes are permitted for bonding grounding electrodes together in industrial and commercial buildings.

Same grounding electrode must be used for all services. Two or more electrodes, bonded together, considered one.

Figure 6-9

Interestingly, no distance between electrodes is given beyond which the electrodes do not have to be bonded together. Buildings of "large area" are permitted by Section 230-2(a) Exception No. 5 to have more than one service. However, nothing in the Code defines the dimensions of a "large building." Some inspection authorities use voltage drop of major feeders for guidance in determining when a building is one of "large area." Where feeder conductors would have to be increased unreasonably to maintain voltage regulation, one or more additional services are permitted.

Section 250-54 requires the grounding electrodes for the multiple services be bonded together no matter how far apart they are in the same building. This is important so there is not more than one earth potential impressed on equipment in or on the building. Section 250-81 requires the bonding jumper(s) used for this purpose to be sized from Table 250-94 and be installed in accordance with Sections 250-92(a) and (b). See Chapter Seven of this text for additional information on installation of grounding electrode conductors.

Section 250-54 also requires that a common grounding electrode be used to ground conductor enclosures and equipment in or on the building and that the same grounding electrode be used to ground the system. This does not mean that one cannot use more than one grounding electrode. But, if more than one is used, then all the grounding electrodes must be bonded together to form a common grounding electrode. Where multiple grounding electrodes are bonded together as cited above, such multiple grounding electrodes become, in effect, a common grounding electrode system.

Earth return prohibited.

No mention is made in the Code to providing a low-resistance, low-impedance, common grounding electrode for clearing ground faults. Reference to Figure 6-10 will show that the grounding electrode is in the earth return circuit. Even if the grounding electrode resistance to earth was very low, it would have little affect on the clearing of a ground fault, because the reactance of the earth return circuit is very high. Where a parallel path through the earth and through a grounded service conductor, about 95 percent of the fault current will return over the grounded service conductor. A low-resistance, common grounding electrode is valuable, however, in holding equipment close to earth potential. It simply is not effective in clearing a line-to-ground fault.

Section 250-51 and 250-91(c) make it clear that grounding electrodes are not permitted to be used in place of equipment grounding conductors. The earth is not to be used as the sole equipment grounding conductor.

If a ground fault should develop as shown in the upper drawing in Figure 6-10, where two separate grounding electrodes are used, the fault current flow will be through the service conductor then through the impedance of the ground fault, the grounding electrode conductor, the grounding electrode, the path through the earth to the grounding electrode at the transformer and finally through the grounding conductor to complete the circuit to the transformer. It would be a rare case where that circuit resistance would add up to less than 12 ohms (while the impedance would be higher). At best, therefore, the fault current would not reach a value high enough to operate a 15 ampere overcurrent device on a 120 volt-to-ground circuit.

Considering resistance only, the shown circuit has two grounding electrodes in series. Compared to the much lower resistance parallel path of the grounded circuit conductor, we have a resistance ratio between the two parallel paths of about 50 times for a No. 2 service, to well over 100 times for the larger services. When we take impedance into account, the ratio will be higher.

Under normal operating conditions we will have some unbalanced current flowing in the neutral. Some unbalanced neutral current will thus flow through the earth, but it will be small in comparison to that which will flow through the grounded service conductor.

Any belief that the circuit to the grounding electrode can be depended on to clear a fault is clearly erroneous no matter how large a grounding electrode conductor we use.

However, when we short-circuit that high-impedance earth path by installing the grounded circuit conductor as shown in the lower drawing in Figure 6-10, we now have a low-impedance path as required in Section 250-51. This will allow a large current to flow over the conductor to allow the overcurrent device to clear the fault and thus provide the safety contemplated by the Code.

If the building is of steel-frame construction, the frame of the building may be used as the grounding electrode if the building frame is effectively connected to earth or "effectively grounded."

Earth Return Prohibited

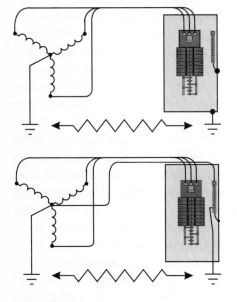

Ground fault at service attempts to return to source (transformer) through the high impedance path through the earth.

A low impedance path for clearing ground faults is provided through the grounded service conductor. (Parallel path through the earth.)

Figure 6-10

Resistance of grounding electrodes.

There is no requirement in Article 250 that the grounding electrode system required by Section 250-81 meet any maximum resistance to ground. No doubt, it is felt that the grounding electrode system will have a resistance to ground of 25 ohms or less.

Maximum Resistance of Grounding Electrodes

- **No maximum for grounding electrode system.**
- **Maximum 25 ohms for "made" electrodes.**
- **Where "made" electrodes exceed 25 ohms - supplement with one additional electrode.**

The rules change for "made" electrodes. The Code states, in Section 250-84, that where a single rod, pipe, or plate electrode does not achieve a resistance to ground of 25 ohms or less, it shall be supplemented by one additional electrode. This means, of course, that where driven ground rods are utilized, two ground rods would be the maximum required under any condition.

Supplemental Electrode

Minimum 6 feet apart or follow manufacturer's instructions.

Made electrode consisting of rod, pipe or plate that exceeds 25 ohms must be augmented by an additional electrode.

Figure 6-11

In general, metallic underground water piping systems, metallic well piping systems, metal frame buildings and similar grounding electrodes may be expected to provide a ground resistance of not over 3 ohms and in some cases as low as 1 ohm.

However, from a practical standpoint, no grounding electrode, no matter how low its resistance, can ever be depended upon to clear a ground fault on any distribution system of less than 1,000 volts.

If a system is effectively grounded as pointed out in the Code under Section 250-51, a path of low impedance (not through the grounding electrode) must be provided to facilitate the operation of the overcurrent devices in the circuit. See Chapter 11 of this text "Clearing Ground-fault Circuits on Distribution Systems."

While the lowest practical resistance of a grounding electrode is desirable and will better limit the voltage to ground when a ground fault occurs, it is more important to provide a low-impedance path to clear a fault promptly, for a voltage to ground can only occur during the period that a fault exists. Clearing a ground fault promptly thus will enhance safety.

In low-voltage distribution systems (600 volts and less) the object is to get the lowest possible grounding electrode resistance. The use of continuous metallic underground water and metal well casings, as well as the metal frames of buildings, should provide a ground resistance not exceeding 3 ohms.

Even though the grounding electrode has low resistance, it is a part of a high-impedance circuit and plays virtually no part in the clearing of a fault on a low-voltage distribution system. This is due in part to being a higher resistance path through the earth than through the grounded service conductor. In addition, the remote path through the grounding electrode and earth is a high-impedance path compared to the circuit where the grounded service conductor is installed and routed with the ungrounded phase conductors.

Objectionable currents.

The Code in Section 250-21 recognizes that conditions may exist which may cause an objectionable flow of current over the grounding electrode conductor other than temporary currents that may be set up under accidental conditions. We should recognize that grounding conductors are not intended to carry current under normal operating conditions. They are installed and intended to carry current to perform some safety function.

The Code does not define what is meant by "objectionable" currents. Clearly, any current over a grounding electrode conductor that would prevent it from maintaining the equipment at the earth potential would be objectionable. Since every conductor has resistance, current flow through the conductor will produce a voltage drop across it. Any voltage drop on a grounding conductor that would create a shock hazard certainly would not be acceptable.

Section 250-21 permits the following corrective actions to be taken:

1. if due to multiple grounds, one or more of such grounds may be abandoned,
2. their location to be changed,
3. the continuity of the grounding conductor may be suitably interrupted, or
4. other means satisfactory to the authority enforcing the Code shall be taken to limit the current over the grounding conductors.

Section 250-21(d) points out that currents that introduce noise or data errors in electronic equipment are not considered to be objectionable currents. Electronic data processing equipment is not permitted to be operated ungrounded or by connection to only its own grounding electrode.

Nature of an earth electrode.[1]

Resistance to current through an earth electrode actually has three components as shown in Figure 6-12.

1. Resistance of the electrode itself and connections to it.
2. Contact resistance between the electrode and the soil adjacent to it.
3. Resistance of the surrounding earth.

Electrode Resistance. Rods, pipes, masses of metal, structures, and other devices are commonly used for earth connections. These are usually of sufficient size or cross-section that their resistance is a negligible part of the total resistance.

Electrode Earth Contact Resistance. This is much less than you might think. If the electrode is free from paint or grease, and the earth is packed firmly, the Bureau of Standards has shown that contact resistance is negligible. Rust on an iron electrode has little or no effect; the iron oxide is readily soaked with water and has less resistance than most soils. But if an iron pipe has rusted through, the part below the break obviously is not effective as a part of the earth electrode.

Resistance of Surrounding Earth. An electrode driven into earth of uniform resistivity radiates current in all directions. Think of the electrode as being surrounded by *shells of earth,* all of equal thickness. See Figure 6-12.

Nature of Earth Electrode

Figure 6-12

The earth shell nearest the electrode naturally has the smallest surface area and offers the greatest resistance. The next earth shell is somewhat larger in area and offers less resistance, and so on. Finally, a distance from the electrode will be reached where inclusion of additional earth shells does not add significantly to the resistance of the earth surrounding the electrode.

Generally, the resistance of the surrounding earth will be the largest of the three components making up the resistance of a ground connection.

The resistance of a made grounding electrode is affected by the resistivity of the soil which in turn is affected by many factors, such as the type of soil, the moisture content of the soil, the salt content of the soil and even the temperature of the soil.

Soil resistivity varies widely from place to place and impacts the grounding electrode resistance. Even in the same area, a variation may be expected from season to season. Soil resistivity may be as low as 500 ohm/centimeters to as much as 100 times or more that figure. The resistance of a grounding electrode or a system of grounding electrodes may be calculated, but the process is complicated, and the formula contains variable factors. Such a calculation serves only as a guide. (An ohm-centimeter [abbreviated ohm-cm] is defined as the resistance of a cube of material [in this case, earth] with the cube sides being measured in centimeters.)

[1] Reprinted with permission of Biddle Instruments, 510 Township Line Road, Blue Bell, PA 19422.

Principles involved in earth-resistance testing.

The resistance to earth of any system of electrodes *theoretically* can be calculated from formulas based upon the general resistance formula:

$$R = p\frac{L}{A}$$

where p is the resistivity of the earth in ohm-cm, L is the length of the conducting path, and A is the cross sectional area of the path. Prof. H. B. Dwight of Massachusetts Institute of Technology developed rather complex formulas for the calculation of the resistance to earth for any distance from various systems of electrodes. All such formulas can be simplified a little by basing them on the assumption that the earth's resistivity is uniform throughout the entire soil volume under consideration.

Because the formulas are complicated, and earth resistivity is neither uniform nor constant, a simple and direct method of measuring earth resistance is needed. This is where an earth resistance tester, a self-contained portable instrument that is reliable and easy to use, is valuable. With it, you can check the resistance of your earth electrode while it's being installed; and, by periodic tests, observe any changes with time.

To understand the principle of earth testing, consider the schematic diagram Figure 6-13(a). Bear in mind our previous observation with reference to the earth shell in Figure 6-12. With increased distance from an electrode, the earth shells are of greater surface area and, therefore, of lower resistance. Now, assume that you have three rods driven into the earth some distance apart and a voltage applied, as shown in Figure 6-13(a). The current between rods 1 and 2 is measured by an ammeter; the potential difference (voltage) between rods 1 and 3 is measured by a voltmeter.

If rod 3 is located at various points between rods 1 and 2, preferably in a straight line, you can get a series of voltage readings. By Ohm's Law (R = E/I) you can determine the earth resistance at any point measured. For example, if the measured voltage (E) between rods 1 and 3 is 30 volts and the measured current (I) is 2 amperes, the resistance of the earth (R) at that point would be 15 ohms.

The series of resistance values can be plotted against distance to obtain a curve. As rod 3 is moved away from rod 1, the resistance values increase, but the amount of increase gets less and less until a point is reached where the rate of increase becomes so small that it can almost be considered

Principles of Earth Resistance Testing

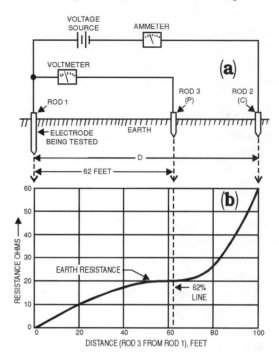

Figure 6-13

constant. The earth shells between the two rods (1 and 3) have so great a surface area that they add little to the total resistance. Beyond this point, as rod 3 approaches the earth shells of rod 2, resistance gradually picks up. Near rod 2, the values rise sharply.

Now, let's say that rod 1 is our earth electrode under test. From a typical earth-resistance curve, what is the resistance to earth of this rod? We call rod 2 current-reference probe C and rod 3, potential-reference probe P (simply for convenience in identification). The correct resistance is usually obtained if P (rod 3) is placed at a distance of about 62 percent of the distance between the earth electrode under test and C (rod 2). For example, in Figure 6-13(b), the distance D from the earth electrode to C is 100 feet. Taking 62 percent of this distance, we get 62 feet. From Figure 6-13(b), the resistance for this distance is 20 ohms. This is the measured resistance of the earth electrode.

This rule works well for simple electrodes such as a driven rod. It also works for a small group of rods. But you must know the true electrical center of the electrode system fairly accurately. Also, accuracy of readings is better if the earth resistivity between the three electrodes is reasonably constant. Finally, C should be far enough away from the earth-electrode system so that the 62 percent distance is out of the "sphere of influence" of the earth-electrode.

Basically, you now have the principle of earth resistance testing. The rest is refinement in test methods, use of electrodes or electrode systems and information about earth resistivity.

How to improve earth resistance.

When you find that your earth-electrode resistance is not low enough, there are several ways you can improve it:

1. Lengthen the earth electrode in the earth.
2. Use multiple rods.
3. Treat the soil.

Effect of Rod Length. As you might suspect, driving a longer rod deeper into the earth materially decreases its resistance. In general, doubling the rod length reduces resistance by about 40 percent. The curve in Figure 6-14 shows this effect. For example, note that a rod driven 2 feet down has a resistance of 88 ohms; the same rod driven 4 feet down has a resistance of about 50 ohms. Using the 40 percent reduction rule, 88 x 0.4 = 35 ohms reduction. A 4-foot deep rod, by this calculation, would have a resistance of 88 minus 35 or 53 ohms comparing closely with the curve values.

Effect of Rod Depth

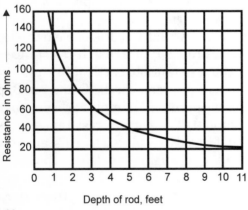

Figure 6-14

You might also think that increasing the electrode diameter would lower the resistance. It does, but only a little. For the same depth, doubling the rod's diameter reduces the resistance only about 10 percent. For example, a 10-foot deep rod $5/_8$ inch in diameter, has a resistance of 6.33 ohms; increasing its diameter to 1¼ inch lowers the resistance only to 5.6 ohms. For this reason, you normally only consider increasing the rod diameter if you have to drive it into hard terrain.

Use of Multiple Rods. Two well-spaced rods driven into the earth provide parallel paths. They are, in effect, two resistances in parallel. The rule for two resistances in parallel does not apply exactly; that is, the resultant resistance is not one-half the individual rod resistances (assuming they are of the same size and depth). Actually, the reduction for two equal-resistance rods is about 40 percent. If three rods are used, the reduction is 60 percent, and if four, 66 percent (see Figure 6-15).

Figure 6-15

When you use multiple rods, they must be spaced further apart than the length of their immersion. There are theoretical reasons for this, but you need only refer to curves such as Figure 6-15. For example, if you have two rods in parallel and 10-foot spacing, resistance is lowered about 40 percent. If the spacing is increased to 20 feet, resistance is reduced to about 50 percent of a single rod.

Treatment of the Soil. Chemical treatment of soil is a good way to improve earth-electrode resistance when you can't drive deeper ground rods, because of hard underlying rock, for example. It is beyond the scope of this manual to recommend the best treatment chemicals for all situations. You have to consider the possible corrosive effect on the electrode. Magnesium sulfate, copper sulfate, and ordinary rock salt are suitable noncorrosive materials. Magnesium sulfate is the least corrosive, but rock salt is cheaper and does the job if applied in a trench dug around the electrode. See Figure 6-16.

Soil Treatment

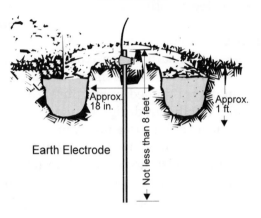

Earth Electrode

Figure 6-16

Chemical treatment is not a permanent way to improve the earth-electrode resistance. The chemicals are gradually washed away by rainfall and natural drainage through the soil. Depending upon the porosity of the soil and the amount of rainfall, the period for replacement varies. It may be several years before another treatment is required.

Chemical treatment also has the advantage of reducing the seasonal variation in resistance that results from periodic wetting and drying out of the soil. However, you should only consider this method when deep or multiple electrodes are not practical.

Grounding electrode study.

The Southern Nevada Chapter of the IAEI has embarked on an ambitious study of the effectiveness of several types of grounding electrodes. The testing program began in 1992. It is planned that the five different sites with 18 electrodes will be monitored over a ten year period. (See the sidebar for a description of electrodes being studied.) Data on electrode resistance, soil moisture, area temperature and rainfall is being collected on a regular basis.

Soil conditions at the five different sites are reported to be as follows:

"Balboa" – predominately gravel rock loosely compacted and is normally dry.

"Lone Mountain" – normally dry with sandy/silty soil.

Description of Electrodes

A. No. 2 AWG stranded copper wire 50' in length centered in 12" of sand at 36" measured to the conductor.

B. No. 4 steel reinforcing bar located within and near the bottom of a concrete foundation consisting of 12" by 12" of concrete 2500 psi. Top of concrete located approx. 6" below grade.

C. No. 4 solid copper wire 25' in length centered in 6" by 6" of Erico ground enhancement material. Located at 20" to the bottom of the concrete.

D. No. 4 solid copper wire 25' in length centered in 6" by 6" of concrete 2500 psi; located at 20" to the bottom of the concrete. This electrode is designed to imitate a thickened edge of a post tensioned concrete construction.

E. 8' by 5/8" copper clad ground rod centered in a pre-drilled 9" dia. by 9' vertical hole, encased by Erico GEM.

F. 8' by 5/8" copper clad ground rod installed horizontally centered in a 6" Erico GEM located 20" to the bottom of the GEM.

G. 8' by 5/8" copper clad ground rod installed horizontally directly buried at 30".

H. 8' by 5/8" copper clad ground rod vertically driven.

I. 10' by 3/4" galvanized ground rod driven vertically.

J. 10' by 3/4" galvanized ground rod directly buried horizontally at a depth of 30".

K. Grounding plate directly buried at 30".

L. Lyncole XIT 10' vertical electrode in pre-drilled 9" hole.

M. Arrangement of steel and concrete designed to imitate a light pole base; six, 2 ft. No. 4 steel vertical rods tied with three No. 2 steel hoops separated 12" vertically in 2' dia. by 30" concrete 2500 psi.

N. No. 4 solid copper wire 20' roll shaped as a coil installed in 2' by 2' concrete 2500 psi.

O. No. 4 solid copper wire 20' roll shaped as a coil installed in 2' by 2' Erico GEM.

P. Wood pole 18" dia. by 8' with pole plate attached at bottom with No. 6 solid copper wire wrapped in a spiral for 6' at 6" spacing between wraps.

Q. 8' by 1/2" copper clad ground rod installed horizontally directly buried at 30".

R. Lightning Eliminators Chemrod vertical installed per manufacturer's recommendations.

S. Lyncole XIT horizontal rod installed per manufacturer's recommendations.

"East Charleston" – generally dry with sand and small rock with surface froth which may be signs of ground salts or sulfides.

"Pecos" – generally dry sand and silt down to a depth of about six feet and has clay and is wet down to about 12 feet.

"Pawnee" – a high water table with silt and clay.

Figure 6-17 gives results of actual on-site resistance measurements made of the various grounding electrodes at each location. The results are an average of the readings taken over the three-year period of 1993 through 1995. For example, electrode "E" (which happens to have the lowest average resistance over the period) has an average resistance of 17.51 ohms at the Balboa site and an average resistance of 2.11 ohms at the Pawnee site. The average for all sites for the three-year period is 6.30 ohms.

The averages are calculated from data furnished. The data is believed to be accurate and is subject to errors and omissions in collection and recording. In some cases, the data is incomplete as indicated for electrodes "R" and "S."

This is an active study and while valuable data is being collected on a regular basis, it may be too early to make conclusions. It appears the grounding electrode study will be expanded to additional sites in different geographical locations which should provide data from a broader perspective.

Lightning rods.

The Code prohibits, in Section 250-86, the use of lightning rod conductors and electrodes used for grounding lightning rods, in place of the grounding electrodes required by the Code for a wiring system and for equipment. In addition, the Code requires that all electrical equipment be kept at least 6 feet from lightning rod conductors or that they be bonded together.

Conclusion.

No potential exists between a conductor or equipment enclosure and earth if the system is properly and adequately grounded, except during a fault. By careful and intelligent design we must reduce the time of clearing a fault to a minimum.

While it is desirable to obtain a grounding electrode having a resistance as low as practical, it is more important that we provide a path of low impedance which will clear a ground fault when it occurs.

No hazard can exist in the distribution system if there is no insulation failure to create a ground fault. Further, the hazard can exist only for the period of time it takes to clear the fault. If a ground fault clears promptly, it is extremely unlikely that any loss of life would occur while property damage would be kept to a minimum.

	Balboa	Lone Mountain	East Charleston	Pecos	Pawnee	Average
Electrode A	35.74	8.26	25.05	18.14	4.19	18.28
Electrode B	33.09	7.27	13.11	11.49	5.36	14.06
Electrode C	29.52	2.94	19.78	8.08	1.64	12.39
Electrode D	36.23	8.90	58.20	27.39	8.09	27.76
Electrode E	17.50	3.57	3.76	4.56	2.11	6.30
Electrode F	32.20	3.24	8.78	4.69	2.42	10.27
Electrode G	163.22	56.93	39.20	18.49	11.55	57.88
Electrode H	44.29	5.35	10.35	4.59	6.19	14.16
Electrode I	35.12	6.03	6.55	7.93	2.89	11.71
Electrode J	191.49	9.23	26.63	11.84	2.89	48.42
Electrode K	424.05	55.52	76.32	138.21	19.93	142.81
Electrode L	18.35	4.73	8.22	5.50	4.78	8.32
Electrode M	71.64	23.57	15.91	21.66	14.24	29.40
Electrode N	76.72	13.52	16.85	31.89	13.96	30.59
Electrode O	63.75	5.39	4.22	26.59	6.05	21.20
Electrode P	72.71	16.11	10.67	6.65	8.84	23.00
Electrode Q	145.04	17.81	43.20	28.70	15.30	50.01
Electrode R	13.57	N/A	N/A	N/A	N/A	13.57
Electrode S	N/A	N/A	N/A	16.87	3.54	10.20

Figure 6-17

For maximum safety, one rule that must be followed is; **ONLY ONE** grounding electrode system must be used in or on a building, with everything that should be grounded connected to that same grounding electrode system.

If more than one grounding electrode is required, they may be used providing **ALL** the grounding electrodes used are bonded together to form a grounding electrode system which in effect is **ONE** common grounding electrode.

Grounding electrodes should never be relied upon to provide the ground-fault return path for equipment. That is not their intended function in the electrical system. They cannot be relied upon to provide that path.

Chapter Six: The questions included here were developed using material included in this chapter. The answers can be found by reviewing the text. It is also important that students make use of the 1996 NEC®, where many answers can be found. See page 236 for answers.

1. Where used to establish a ground for connecting electrical systems to the earth this conductor is defined as the ____.
   a. equipment grounding
   b. grounding electrode
   c. main bonding jumper
   d. earthing conductor

2. Intentionally connected to earth through a ground connection or connections of sufficiently low impedance and having sufficient current-carrying capacity that will prevent buildup of voltages that may result in undue hazard to connected equipment or to persons, defines ____.
   a. effectively grounded
   b. sufficiently grounded
   c. approved
   d. identified

3. One or more steel reinforcing bars or rods not less than $\frac{1}{2}$ inch in diameter, or 20 feet or more of bare copper conductor not smaller than No. 4 is defined as a ____.
   a. butt ground
   b. pole ground
   c. feeder electrode
   d. concrete-encased electrode

4. A concrete-encased electrode must be located within or near the bottom of a foundation or footing and is required to be encased by not less than ____ inches of concrete.
   a. 0
   b. 1
   c. 2
   d. 6

5. A copper conductor not smaller than No. 2, at least 20 feet long encircling a building or structure, and buried at not less than 2½ feet deep defines a ____.
   a. encased electrode
   b. ground ring
   c. Ufer ground
   d. common bonding grid

6. Where an underground metal water pipe is the only grounding electrode that is available, and is connected at the building or structure served, it must be supplemented by another ____.
   a. grounding electrode
   b. main bonding jumper
   c. equipment grounding conductor
   d. circuit bonding jumper

7. An electrode that is considered as being suitable for supplementing the metal underground water pipe, includes a concrete-encased electrode, ground ring, other local metal underground systems or structures, or ____ electrodes.
   a. rod
   b. pipe
   c. plate
   d. all of the above

8. If available on the premises, at each building or structure served, all grounding electrodes, including ____ electrodes, are required to be bonded together to form the grounding electrode system.
   a. identified
   b. approved
   c. made
   d. gas pipe

9. Where the electrodes described in Section 250-81 are not available at the service location, a grounding electrode must be made. It may be a ____.
   a. local metallic underground piping system, tanks, etc.
   b. pipe or conduit electrodes not less than 8 feet in length nor smaller than $\frac{5}{8}$ inch, and if of iron or steel, must be galvanized or metal-coated for corrosion protection.
   c. rod electrodes of steel or iron at least $\frac{5}{8}$ inch diameter.
   d. any of the above

10. Where two or more grounding electrodes are bonded together, they are considered to be ____ electrode.
   a. one
   b. an identified
   c. an approved
   d. a listed

11. The Code recognizes that conditions may exist that may cause an objectionable flow of current over the grounding electrode conductor, other than temporary currents that may be set up under accidental conditions. Permitted alterations include ____.
    a. if due to multiple grounds, one or more, but not all, of such grounds may be abandoned
    b. their location can be changed
    c. continuity of the grounding conductor may be suitably interrupted
    d. any of the above

12. Currents that introduce noise or data errors in electronic equipment are not considered to be ____.
    a. dangerous currents
    b. objectionable currents
    c. harmonic currents
    d. unsafe currents

13. Where a single rod, pipe, or plate electrode does not achieve a resistance to ground of ____ ohms or less, it is required to be supplemented by only one additional electrode.
    a. 50
    b. 40
    c. 30
    d. 25

14. The use of continuous metallic underground water and metal well casings, as well as the metal frames of buildings, should provide a ground resistance not exceeding ____ ohms.
    a. 3
    b. 6
    c. 12
    d. 25

15. In general, doubling the length of a rod type grounding electrode reduces the resistance by about ____ percent.
    a. 10
    b. 20
    c. 30
    d. 40

16. The use of lightning rod conductors and electrodes are ____ for grounding lightning rods in place of the grounding electrodes required by the Code for a wiring system and for equipment.
    a. permitted to be used
    b. prohibited from being used
    c. permitted by special permission
    d. identified when labeled to be used

17. The Code requires that all electrical equipment be kept at least ____ feet from lightning rod conductors or that they be bonded together at the point where the separation is ____ feet or less.
    a. 3
    b. 6
    c. 8
    d. 12

Chapter **7**

Grounding Electrode Conductors

Objectives

After studying this chapter, the reader will be able to understand

- General requirements and definitions for grounding electrode conductors.
- Functions of the grounding electrode conductor.
- Grounding electrode conductor connections.
- Material and protection for grounding electrode conductors.

Definition.

Grounding electrode conductor: "The conductor used to connect the grounding electrode to the equipment grounding conductor, to the grounded conductor, or to both, of the circuit at the service equipment or at the source of a separately derived system."*

General requirements.

The grounding electrode conductor is the conductor used to connect the wiring system, or the equipment grounding conductor, or both, to a grounding electrode, hence the specific term "Grounding electrode conductor." It must be properly sized based on the size of the service conductors and is not required to exceed No. 3/0 copper or 250 kcmil aluminum. Specific requirements are given regarding conductor material, how it is to be installed so as to be secure from damage and how it must be connected.

Function of grounding electrode conductor.

The grounding electrode conductor is the sole connection between the grounding electrode and the grounded system conductor and equipment grounding conductor for a grounded system, or the sole connection between the grounding electrode and the service equipment enclosure and equip-

Grounding Electrode Conductor - Grounded System

Grounded system conductor

Equipment grounding conductor

Grounding electrode conductor connects system and equipment to the grounding electrode.

Grounding electrode (system)

Figure 7-1

Grounding Electrode Conductor - Ungrounded System

Equipment grounding conductor

Grounding electrode conductor connects equipment to the grounding electrode.

Grounding electrode (system)

Figure 7-2

ment grounding conductor for an ungrounded system. [Section 250-53(a)]

A common grounding electrode conductor is required to ground both the circuit grounded conductor and the equipment grounding conductor. In other words, one grounding electrode conductor cannot be used to ground the circuit conductor and another grounding electrode conductor used to ground the equipment grounding conductor even though both grounding conductors are connected to the same grounding electrode. A single grounding electrode conductor must be used for both the circuit and the equipment.

Maximum current.

In all grounded systems of 600 volts or less, the maximum current that will flow in the grounding electrode conductor is dependent on the sum of the resistance of the grounding electrode at the service, the grounding electrode at the transformer bank plus the resistance of the earth path between the two grounding electrodes. In addition, there will be the resistance of the grounding electrode conductors themselves.

As illustrated in Figure 7-3, if we assume that the sum of these resistances is equal to 22 ohms (a figure that may be lower than is found in actual practice), then, in a 120 volt-to-ground system, the maximum current through the grounding electrode conductor and grounding electrodes will be 5.5 amperes (120 volts ÷ 22 ohms = 5.5 amperes). As can be seen, this grounding connection is ineffective. Equipment grounded in this manner is

High Impedance Earth Return

Figure 7-3

unsafe as a ground fault would not cause an up-stream overcurrent protective device to clear the fault from the equipment. Voltage levels that could be fatal could appear on equipment that is grounded in this manner.

This maximum current value is arrived at by considering resistance only. If the reactance of the circuit is included, the net result will be an impedance value much higher than the resistance value. The actual current that will flow in the grounding electrode conductor will thus be considerably less than the figures as calculated based on resistance only. Ground-fault currents through the grounding electrode conductor are thus not liable to attain a value anywhere near the continuous rating of the conductor even under ground-fault conditions.

Sizing the grounding electrode conductor at a single service.

The grounding electrode conductor must be sized in accordance with Table 250-94 of the Code. That conductor is required to be a minimum size of No. 8 and need not be larger than No. 3/0 copper.

Sizing Grounding Electrode Conductors

- **Refer to Table 250-94.**

- **Go down left column and find size of service-entrance conductor used.**

- **Go across to find size of copper or aluminium grounding electrode conductor.**

Where aluminum or copper-clad aluminum grounding electrode conductors are installed they are required to be not smaller than No. 6 nor larger than 250 kcmil. The size of this grounding electrode conductor is based upon the size of the service entrance conductor by reference to Table 250-94. Table 250-94 is reprinted in Chapter 18 of this text.

For example, where a 3/0 copper service-entrance conductor is installed, the minimum size grounding electrode conductor is No. 4 copper or No. 2 aluminum. If 750 kcmil aluminum service entrance conductors are installed, a 1/0 copper or 3/0 aluminum grounding electrode conductor must be used.

Sizing grounding electrode conductor for service with parallel conductors.

Where service-entrance conductors are installed in parallel as allowed by Section 310-4, the circular mil area of the conductors added together and treated as a single conductor for purposes of sizing the grounding electrode conductor. For example, if four 250 kcmil copper conductors are installed, they are considered to be a single 1,000 kcmil conductor. In this case, by reference to Table 250-94, we find the minimum grounding electrode conductor to be a 2/0 copper or 4/0 aluminum conductor.

Grounding Electrode Conductor for Parallel Service-Entrance Conductors

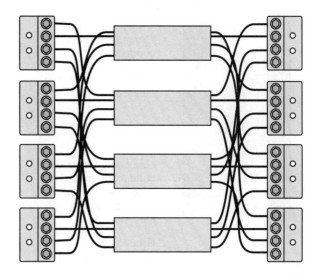

Four 250 kcmil aluminum conductors per set
4 x 250 = 1000 kcmil
Refer to Table 250-94
Select 2/0 copper or 4/0 aluminum GEC
See exceptions for GEC to made electrodes.

Figure 7-4

If No. 1/0 through 4/0 AWG conductors are installed in parallel, they must first be converted to circular mil area before applying Table 250-94. For example, if 2/0 aluminum service-entrance conductors are installed in parallel, refer to NEC® Table 8 of Chapter 9 for determining the conductor's circular mil area. (NEC® Table 8 of Chapter 9 is reprinted in Chapter 18 of this text.) There, we find it to have an area of:

133,100 circular mils;

133,100 cm x 2 = 266,200 cm.

By reference to Table 250-94 we find that a No. 2 copper or 1/0 aluminum grounding electrode conductor is required.

Sizing grounding electrode conductors for multiple enclosure services.

Services are permitted to be installed in up to six enclosures installed at one location or at separate locations. The method of sizing grounding electrode conductors is a matter of choice. One basic rule must be followed - size the grounding electrode conductor for the circular mil area of service-entrance conductor(s) at the point of connection, or for the size of the main service-entrance conductor. (Section 250-91 Exception No. 2)

GEC Connection to Electrode

Individual GEC to grounding electrode

Clamp approved for two wires or two clamps

Figure 7-6

service disconnecting means enclosure to the grounding electrode. In this case, the grounding electrode conductor is sized for the service-entrance conductor serving that enclosure. See Figure 7-6.

GEC Connection in Wireway

Grounded service conductor

500 kcmil ungrounded service entrance conductor

1/0 copper 3/0 aluminum Grounding electrode conductor

Figure 7-5

In Figure 7-5 with assumed 500 kcmil copper service-entrance conductors, the service-entrance grounded conductor is shown grounded inside the wireway. By reference to Table 250-94, we find that the grounding electrode conductor must be 1/0 copper or 3/0 aluminum.

If the installer chooses to, an individual grounding electrode conductor may be installed from each

Taps to Main Grounding Electrode Conductor

GEC Tap →

Main GEC →

Grounding Electrode →

Figure 7-7

In addition, a single grounding electrode conductor is permitted to serve several enclosures. The **main** grounding electrode conductor is sized for the main service-entrance conductors. Taps that are sized for the individual service-entrance conductors are connected from the service to the main grounding electrode.

In Figures 7-7 and 7-8, it is assumed that No. 2 copper conductors serve each service disconnecting means from the wireway. Shown are taps to a **main** grounding electrode conductor. The tap conductors must be connected to the main grounding electrode conductor in such a manner that it is not spliced. This means that the tap conductor must be connected with a split bolt or tap connector allowing the **main** grounding electrode conductor to pass through the connector unbroken. The concept of tapping the grounding electrode conductor applies whether the sets of service-entrance conductors are tapped from a gutter or are installed individually as an overhead or underground system.

Tapping Grounding Electrode Conductor

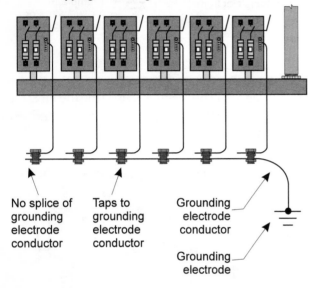

No splice of grounding electrode conductor

Taps to grounding electrode conductor

Grounding electrode conductor

Grounding electrode conductor

Grounding electrode

Figure 7-8

For Figure 7-8, the size of the **main** grounding electrode conductor is determined as follows.

No. 2 conductor = 66,360 circular mils
(Table 8, Chapter 9)
6 x 66,360 circular mils = 398,160 circular mils

From Table 250-94 we find the **main** grounding electrode conductor to be 1/0 copper or 3/0 aluminum since 398 kcmil is between 350 and 600 kcmil.

The tap conductors from the individual enclosures are sized from Table 250-94 and are No. 8 copper or No. 6 aluminum.

If the installer chose to, the grounding electrode conductors from the service disconnects may be connected individually to the grounding electrode rather than being tapped to the main grounding electrode conductor.

Exceptions to size of grounding electrode conductor.

A three-part exception is provided to the minimum sizes of grounding electrode conductors specified in Table 250-94.

The first exception allows the grounding electrode conductor to be not larger than No. 6 copper or No. 4 aluminum wire where it is the sole connection to a rod, pipe or plate electrode. The second exception provides that the grounding electrode conductor that is the sole connection to a concrete encased grounding electrode need not be larger than No. 4 copper wire. Note that aluminum wire is not permitted for this application.

The third part of the exception provides that, where connected to a ground ring, the grounding electrode conductor need not be larger than the ground ring conductor. The minimum size and conductor material for a ground ring is No. 2 copper. That requirement assumes, and correctly so, that the resistance of such electrodes will limit the maximum current that may be expected to flow through them.

The Code states that the grounding conductor need not be larger than No. 6 copper wire or No. 4 aluminum where connected to a "made" electrode. The use of this smaller No. 6 conductor has its basis on the fact that the No. 6 conductor can never be called upon, even under ideal conditions, to carry a current beyond its safe short-time rated capacity.

Grounding electrode conductor connections.

The Code requires that the point of connection of the grounding conductor to grounding electrodes shall be accessible and made in a manner that will assure a permanent and effective ground. An exception provides that a connection at a concrete-encased, driven, or buried grounding electrode is not required to be accessible. (Section 250-112)

Grounding electrode conductor connections must be by exothermic welding, listed lugs, listed pressure connectors, listed clamps or other listed means. The only connection means not required to be listed are those made by exothermic welding. Connections depending solely on solder shall not be used.

Under the heading "Connection to Electrodes" in Section 250-112, the Code requires that not more than one conductor shall be connected to the grounding electrode by a single clamp or fitting, unless the clamp or fitting is listed as being suitable for connecting multiple conductors.

Grounding Electrode Conductor Connections

Connection must be accessible

Connection to concrete-encased driven or buried electrodes not required to be accessible

Figure 7-9

Ground clamps must be listed for both the materials of the grounding electrode and the grounding electrode conductor. For example, ground clamps must be listed for aluminum conductors to be used for such connections. Clamps used on a pipe, rod or other buried electrode must be listed for direct earth burial. (Section 250-115)

Ground Clamps Listed for Application

- Grounding conductor connected to electrode by exothermic welding, listed lugs, listed pressure connectors, listed clamps or other listed means.

- Ground clamps shall be listed for materials of grounding electrode and grounding electrode conductor.

- Shall be listed for direct soil burial where used on pipe, rod or other buried electrodes.

Figure 7-10

The Code does not permit the use of a sheet-metal-strap type of ground clamp unless it is listed and attached to a rigid metal base. Section 250-115(c).

Also, Underwriters Laboratories Guide Card information (KDAX) states, "Strap type ground clamps are not suitable for attachment of the grounding conductor of an interior wiring system to a grounding electrode." Use strap type ground clamps for only the type of conductor and grounding electrode, as well as the environment it is listed and labeled for, such as for communication circuits.

Protection of Ground Clamp Attachment

Ground clamp to be approved for use without protection or be protected from "ordinary" damage unless it is:
(a) Not likely to be damaged, or it is
(b) Protected in metal, wood or equal.

Figure 7-11

It is required that ground clamps be protected from physical damage unless the fitting is one that is approved for use without protection. Section 250-117. Where installed for direct soil burial, the clamp must be listed for such use. Section 250-115.

Clean surfaces.

It is require in Section 250-118 that nonconductive coatings such as paint, lacquer and enamel be removed from threads and other contact surfaces of equipment to be grounded or be connected by means of fittings designed to make such removal unnecessary. (Underwriters Laboratories reported in a June 27, 1995 letter that no fittings listed by them incorporate this feature.) This will ensure a good electrical connection. See Chapter 8 of this text for additional discussion on the subject.

Here, again, it is important to consider that all grounding conductors and their connections form a part of a circuit which is required under certain conditions to carry not only a current but one that may be many times the full-load rating of the conductor involved if the conductor was being used for continuous duty in an electrical circuit.

Grounding electrode conductor material.

The grounding electrode conductor is required to be of copper, aluminum or copper-clad aluminum and may be solid, stranded, insulated, covered or bare. The material selected shall be resistant to any corrosive condition it may be exposed to. As an option, it may be suitably protected against corrosion. Section 250-91(a).

The required current-carrying capacity of the grounding electrode conductor is limited by the resistance of the grounding electrode and the earth return path. In a grounded system, it is further limited by the fact that it completes a parallel circuit from the grounded conductor through both grounding electrodes. The other side of the parallel circuit is the grounded service conductor which is a circuit of much lower impedance.

Installation of grounding electrode conductors.

Section 250-92(a) covers the installation requirements of grounding electrode conductors. This section requires that the grounding electrode conductor or its enclosure be securely fastened to the surface on which it is carried.

Protection of Grounding Electrode Conductors

- **GEC or enclosure to be securely fastened to surface.**

- **No. 4 or larger—protect from severe physical damage.**

- **No. 6 run along surface and be securely fastened or be protected.**

- **Smaller than No. 6—protect from damage.**

- **Aluminum—not allowed in contact with masonry, the earth, where subject to corrosive conditions or within 18 inches of the earth.**

No. 4 or larger conductors require protection where exposed to "severe physical damage" and thus leaves it to a judgment call on the part of the inspector as to what constitutes "severe physical damage." No. 6 conductors must be run along the surface of the building construction and be securely fastened. Otherwise, they must be protected by installation in rigid or intermediate metal conduit, rigid nonmetallic conduit, electrical metallic tubing or cable armor. Grounding conductors smaller than No. 6 must be protected in one of the methods allowed for No. 6 conductors.

In addition, aluminum and copper-clad aluminum conductors are not allowed to be used in direct contact with masonry or the earth, or where subject to corrosive conditions. Aluminum or copper-clad aluminum conductors are not permitted to be installed outside within 18 inches of the earth. They also cannot be used for connection to concrete encased electrodes. Aluminum conductors are otherwise permitted to be used as grounding electrode conductors where the clamp or connector are listed for both the conductor material and the electrode.

Splicing grounding electrode conductor.

The general rule in Section 250-91(a) is that grounding electrode conductors are to be installed without a splice or joint. Exception No. 1 permits splices in busbars as they are at times used as a grounding electrode conductor. Exception No. 2 permits extending a tap from the main grounding electrode conductor into each service disconnecting means where more than one service disconnecting means is installed as permitted in Section 230-40, Exception No. 2. (Covered earlier in this chapter.)

Splicing Grounding Electrode Conductor

Figure 7-12

Exception No. 3 to Section 250-91(a) permits the grounding electrode conductor to be spliced by two methods; irreversible compression-type connectors listed for the purpose or by the exothermic welding process. It is vital that the manufacturer's instructions be carefully followed where either of these splicing methods are chosen.

Where compression-type connectors are used, the correct splicing sleeve for the conductor material and size must be selected. Then, the compression tool, and in some cases the proper die for the sleeve to be crimped, must be used. These compression-type connectors must be specifically listed by a qualified electrical testing laboratory for the purpose of splicing grounding electrode conductors.

Where exothermic welding of the grounding electrode is performed, the correctly sized form or mold for the conductor to be spliced must be used. Also, unless specifically permitted otherwise by the manufacturer, the conductors to be spliced by this method must be clean and dry. Inspect the resulting splice carefully to be certain that it has been made satisfactorily.

Protecting grounding electrode conductor from magnetic field.

Where metal enclosures are provided for protection of the grounding electrode conductor, some special procedures must be followed. This is required by Section 250-92(b) and 250-71(a)(3).

Protecting Grounding Electrode Conductor From Magnetic Field

Metal conduit nipples (typical)

Bonding at each end and to all intervening raceways, boxes and enclosures between the service equipment and the grounding electrode. (Neutral connections not all shown.)

Metal conduit enclosing grounding electrode conductor

Grounding electrode

Figure 7-13

1. The metal conduit must be electrically continuous from the point of attachment to cabinets or equipment to the grounding electrode.
2. The conduit must be securely fastened to the ground clamp or fitting.
3. The metal conduit that is not physically continuous from the enclosure to the ground clamp must be made continuous by bonding both ends of the conduit to the conductor.
4. Bonding of the metal raceway or armor that encloses a grounding electrode must be done at each end and to all intervening raceways,

boxes and enclosures between the service equipment and the grounding electrode. Section 250-71(a)(3).

It is common practice to use a No. 8 grounding electrode conductor protected by metallic armored cable. The need for bonding the metallic armor of such cable as required in Section 250-71(a)(3). Where that bonding procedure is not followed, the impedance of the grounding electrode conductor is doubled with the result that its effectiveness is markedly reduced.

Impedance of conduit and conductor.

Table Two of Chapter 18 of this text compares the continuous rating of copper grounding electrode conductors with the service conductors with which they must be used.

Table 7-14. Division of current in conduit.				
Conductor	Conduit Inches	Total Current Amperes	Current in Conductor, Amperes	Current in Conduit, Amperes
6	½	100	3	97
6	½	300	5	295
2	¾	90	7	83
2	¾	350	10	340
2/0	1	150	15	135
2/0	1	590	5	585
4/0	1¼	225	15	210
4/0	1¼	885	15	870

The above test data confirm that, for all practical purposes, the impedance of a conductor enclosed in a steel conduit, when the conduit is bonded to the conductor at both ends, is approximately equal to the impedance of the conduit.

Table Three of Chapter 18 of this text compares the resistance and impedance of copper grounding electrode conductors where enclosed in a steel conduit for physical protection. The last two columns of that table show how the impedance of the conductor is approximately doubled where the conduit is not properly installed as required in Section 250-92. The Code requires that the conduit be bonded at both ends of the grounding electrode conductor to form a parallel circuit with the copper grounding electrode conductor. That important rule, if not observed, will result in doubling the impedance of the grounding conductor. Test data to confirm the above are available. Of course, where the impedance of the installation is increased, the effectiveness is reduced.

Where grounding electrode conductors are used on an alternating-current circuit and enclosed in conduit it is necessary to compare the impedance values and not the resistance values.

The data in Table Three of Chapter 18 of this text shows that where a No. 8 copper conductor is installed in steel conduit and properly bonded at each end, the impedance of the circuit is about the same as where the No. 8 copper conductor was used alone.

For all other sizes of copper conductors installed in the proper sized conduit, the impedance of the circuit is greater where the conduit is used, as compared to using the copper conductor alone. The impedance values are from about 40 percent more for a No. 6 copper conductor in a ¾-inch conduit to about 500 percent more for a No. 3/0 copper conductor in a 1¼-inch conduit, as compared to not using a conduit for physical protection.

Over 100-foot grounding electrode conductor.

The short-time rating of a copper conductor is related to the I^2t rating of the conductor for a given temperature rise which will not damage adjacent insulated conductors or affect the continuity established by the bolted joints. For a period of 5 seconds the short-time rating may be taken as approximately 1 ampere for every 42.25 circular mils area. A No. 6 conductor has an area of 26,240 circular mils and is thus capable of carrying about 621 amperes for 5 seconds safely.

Based on the safe I^2t values for the circuit comprising the various grounding electrode conductors, it will be seen that for a 5 second flow of current, the IR drop in the different sizes of grounding electrode conductors will be approximately 37 volts per 100 feet. Using that figure as a standard, it is recommended that where a grounding electrode conductor exceeds 100 feet in length, the conductor cross section be increased to keep the IR drop to not over 40 volts when carrying the maximum short-time current for the size conductor specified for 5 seconds (I^2t value). The National Electrical Code® does not, now, place a limit on the length of the grounding electrode conductor.

An example of selecting the proper size grounding electrode conductor for a run exceeding 100 feet is as follows.

Given: a No. 1/0 copper service-entrance conductor. The grounding electrode conductor specified in Table 250-94 is a No. 6 copper and the length of the grounding electrode conductor is 150 feet.

If a No. 6 conductor is used which has 26,240 circular mils, resulting in a short-time rating of 621 amperes, and a DC resistance of 0.0737 ohms for 150 feet (0.491 ohms/M ft.), the voltage drop would be 621 x 0.0737 or 46 volts.

We must thus select a larger grounding electrode conductor whose resistance times the short-time rating of the No. 6 conductor in amperes must not exceed 40 volts.

The next larger-sized grounding conductor a No. 4, has a resistance of 0.0462 ohms for 150 feet (0.308 ohms/M ft.), so the voltage drop would be 621 x 0.0462 or 28.7 volts. That would make a No. 4 copper grounding conductor the proper size to use for a service using a No. 1 or No. 1/0 copper service-entrance conductor and having a grounding electrode conductor run of 150 feet.

Direct current systems.

Where used on a direct-current circuit, we do not destroy the value of the grounding conductor if the conduit is properly installed, that is, bonded at both ends to the grounding conductor. The resultant resistance is lower where the conduit is used for physical protection. However, that assumes a steady direct current flow. Special considerations must be given if a direct-current circuit is to be properly protected with a grounding electrode conductor against transient currents such as are produced by lightning. It is necessary to treat the selection of the grounding electrode conductor as would be done for an alternating-current circuit.

Table Four of Chapter 18 of this text shows that where an aluminum conduit is used to enclose an aluminum grounding electrode conductor, the required conduit size has a lower resistance than the aluminum grounding electrode conductor in every case. In the case of aluminum wire size No. 6, the conduit is about one-tenth the resistance of the conductor. For the largest aluminum grounding electrode conductor size, 250 kcmil, the conduit is about half the resistance of the conductor. Thus, where an aluminum grounding electrode conductor is protected with an aluminum conduit and properly bonded at both ends, we will always have a much lower impedance than where the wire is not installed in an aluminum conduit. However, aluminum is subject to certain restrictions owing to chemical corrosion.

Since aluminum conduit is nonmagnetic and has lower resistance as well as impedance values compared to steel, the use of aluminum conduit for physical protection will provide lower impedance values.

Conclusion.

All of the above means that we decrease the safety of an installation (which requires a grounding electrode conductor larger than No. 8) where we enclose the conductor in a steel conduit. Although we decrease physical damage to assure the integrity of the grounding electrode conductor, another hazard is introduced by decreasing the effectiveness of the grounding electrode conductor through increasing the voltage rise on it. No better case could thus be made for not permitting the use of a steel conduit on a grounding electrode conductor larger than a No. 8.

The obvious solution, where physical protection is necessary, is to use nonmetallic or aluminum conduit for enclosing the grounding electrode conductor. That is especially true in view of the improvement in the art of manufacturing nonmetallic conduit which now can be obtained in ample physical strength to meet the requirements of proper physical protection.

Chapter Seven: The questions included here were developed using material included in this chapter. The answers can be found by reviewing the text. It is also important that students make use of the 1996 NEC®, where many answers can be found. See page 236 for answers.

1. A conductor used to connect the grounding electrode to the equipment grounding conductor and/or to the grounded conductor of the circuit at the service equipment, or at the source of a separately derived system is defined as a ____.
   a. main bonding jumper
   b. grounding electrode conductor
   c. feeder bonding jumper
   d. grounded

2. A grounding electrode conductor must be properly sized and is based on the size or rating of the:
   a. service breaker or fuse
   b. transformer
   c. service conductors
   d. grounding electrode

3. For service conductors sized at over 1100 kcmil copper or 1750 kcmil aluminum or copper-clad aluminum, the grounding electrode conductor must be a ____ copper or ____ aluminum.
   a. No. 2/0 - No. 4/0
   b. No. 1/0 - No. 3/0
   c. No. 3/0 - 250 kcmil
   d. No. 3/0 - No. 4/0

3. In all grounded systems of ____ volts or less, the maximum current that will flow in the grounding electrode conductor is dependent on the sum of the resistance of the grounding electrode at the service, the grounding electrode at the transformer bank, plus the resistance of the earth path between the two grounding electrodes.
   a. 600
   b. 1,000
   c. 700
   d. 800

4. Where of copper, the grounding electrode conductor is required to be sized at not less than No. ____.
   a. 10
   b. 14
   c. 8
   d. 12

5. Where a 3/0 copper service-entrance conductor is installed, the minimum size grounding electrode conductor is No. ____ copper or No. ____ aluminum, or copper-clad aluminum.
   a. 8 - 4
   b. 6 - 6
   c. 4 - 2
   d. 4 - 4

6. Where four 250 kcmil copper service entrance conductors are installed per ____, they are considered to be a single 1,000 kcmil conductor.
   a. service
   b. feeder
   c. cable
   d. phase

7. The minimum size for a grounding electrode conductor for a 1,000 kcmil service entrance conductor is No. ____ copper or No. ____ aluminum, or copper-clad aluminum conductor.
   a. 1/0 - 2/0
   b. 3/0 - 4/0
   c. 1/0 - 3/0
   d. 2/0 - 4/0

8. Services are permitted to be installed in up to ____ enclosures where they are installed at one location or at separate locations.
   a. one
   b. two
   c. six
   d. eight

9. A grounding electrode conductor is required to be sized at not less than No. ____ copper or No. ____ aluminum wire where it is the sole connection to a rod, pipe, or plate electrode.
   a. 8 - 8
   b. 6 - 4
   c. 6 - 6
   d. 4 - 8

10. A grounding electrode conductor that is the sole connection to a concrete-encased grounding electrode is not required to be larger than No. ____ copper wire.
   a. 6
   b. 4
   c. 8
   d. 10

11. Grounding electrode conductor connections are required to be by ____, listed clamps, or other listed means.
   a. exothermic welding
   b. listed lugs
   c. listed pressure connectors
   d. any of the above

12. Which one of the following statements is INCOR-RECT?
   a. Exothermic welding connections are not required to be listed.
   b. Soldered connections are permitted for connections to a grounding electrode.
   c. Unless listed, not more than one conductor can be connected to the grounding electrode by a single clamp or fitting.
   d. Clamps used on a pipe, rod, or other buried electrode must also be listed for direct earth burial.

13. Where exposed to severe physical damage, No. ____ or larger grounding electrode conductors require protection.
   a. 4
   b. 8
   c. 6
   d. 3

14. No. ____ grounding electrode conductors must be run along the surface of the building construction and be securely fastened, or they must be protected by installation in rigid or intermediate metal conduit, rigid nonmetallic conduit, electrical metallic tubing, or cable armor.
   a. 8
   b. 6
   c. 4
   d. 2

15. Aluminum or copper-clad aluminum conductors are not permitted to be installed outside within ____ inches of the earth.
   a. 12
   b. 18
   c. 14
   d. 16

Chapter **8**

Bonding Enclosures and Equipment

Objectives
After studying this chapter, the reader will be able to understand
- Requirements for maintaining continuity of ground-fault path.
- Bonding of systems over 250 volts to ground.
- Bonding multiple raceway systems.
- Bonding of receptacles.
- Bonding of metal piping systems.
- Bonding of exposed structural steel.

Maintaining Continuity.

The requirement is given in Section 250-75 for maintaining continuity, by bonding if necessary, at metal raceways, cable trays, cable sheaths, enclosures, frames, fittings and other metal noncurrent-carrying parts used as equipment grounding conductors. This may be necessary to assure that these systems have electrical continuity and the current-carrying capacity to safely conduct the fault current likely to be imposed on them. Note that this bonding of raceways, etc. must be performed regardless of whether or not an equipment grounding conductor is run with the raceway, etc. This will ensure that the raceway will not become energized by a line-to-enclosure fault without having the capacity and capability of clearing the fault by allowing sufficient current to flow to operate the overcurrent protective device.

Bonding to Maintain Continuity

Figure 8-1

Keep in mind that the "weakest link rule" applies to the ground-fault return path. To provide adequate safety, the ground-fault return path must be (1) permanent and electrically continuous, (2) have the capacity to conduct safely any fault current likely to be imposed on it, and (3) have sufficiently low impedance to limit the voltage to ground and to facilitate the operation of the circuit protective devices. (Section 250-51) This ground-fault path must meet all three conditions from the farthest enclosure or equipment all the way back to the service equipment and ultimately to the source. This path can be through many boxes, conduit or other raceways, pull boxes, wireways, auxiliary gutters, panelboards, motor control centers and switchboards. Every connection is important. It only takes one loose or broken fitting to break a link in the fault-current chain.

Section 250-75 also refers to conditions where a nonconducting coating might interrupt the required continuity of the ground-fault path, and it points out that such coatings must be removed unless the fitting(s) is(are) designed as to make such removal unnecessary. Underwriters Laboratories confirmed in writing that, as of June 27, 1995, no fittings having such a feature were listed by them.

Testing of conduit fittings.

The importance of removing paint from enclosures where the conduit or raceway is intended to serve as the fault-current path is further emphasized in a report on "Conduit Fitting Ground-Fault Current Withstand Capability" issued by Underwriters Laboratories on June 1, 1992. Over 300 conduit fitting assemblies from ten different manufacturers were subjected to a current test to simulate performance under ground-fault conditions.

A sample assembly consisted of a conduit fitting secured to one end of a two-foot length of conduit and attached to a metal enclosure.

After securing the conduit fitting to the conduit properly, the conduit fitting was secured to the enclosure using the locknut provided by the manufacturer. The locknut was first hand-tightened and then further tightened ¼ turn with a hammer and standard screwdriver. The fittings were installed through holes in the enclosures that were punched rather than being installed in preexisting knockouts. A pipe clamp, wire connector, conductors and a power supply were assembled to complete the testing. Thermocouples were placed at strategic locations to record pertinent temperature data. Figure 8-2 is a drawing of the sample assembly.

Sample Assembly

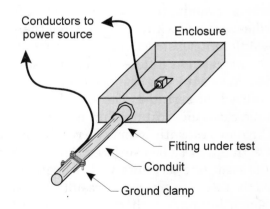

Figure 8-2

Fittings for conduit in the 3/8-inch through 3-inch trade sizes were tested. The appropriate current applied to the fittings in the test program is as shown in the following table.

TEST CURRENTS AND TIMES

Conduit Trade Size, Inch	Test Time	Current, Amperes
3/8	4 Sec.	470
1/2	4 Sec.	1180
3/4, 1	6 Sec.	1530
1-1/4, 1-1/2	6 Sec.	2450
2	6 Sec.	3900
2-1/2	6 Sec.	4900
3, 3-1/2, 4	9 Sec.	5050
4-1/2	9 Sec.	6400
5, 6	9 Sec.	8030

This test should not be confused with a short-circuit withstand test and is not intended to test the maximum short-circuit current these fittings can withstand. Due to the time and current involved, a great deal of heat is generated in the test assembly.

Seven of the more than 300 assemblies tested sustained damage. A visual examination of sample assemblies that failed showed that melting of the die cast zinc locknuts occurred as a result of the fault current. Melting of the die-cast zinc body occurred on five sample assemblies. The painted enclosures on which the fittings were tested were also examined. The examination indicated that melting of the die-cast zinc was probably due to the inability of the locknut to penetrate through the enclosure paint and provide good electrical contact between the fitting and metal of the enclosure.

A visual examination of all the conduit fittings with die-cast zinc locknuts showed that there were three different constructions of the locknuts. The three constructions differed in that the surface of the locknut contacting the enclosure was either flat, nibbed, or serrated. The sample assemblies with die-cast zinc locknuts that did not complete the Current Test with acceptable results had locknuts with flat or ribbed surfaces. All fittings having die cast zinc locknuts with serrations completed the test with acceptable results. It appeared as though locknuts with serrations consistently penetrated through the enclosure paint and provided better electrical contact between the fitting

Figure 8-3. Sample showing signs of arcing at conduit fitting connection to enclosure. (Photo courtesy Underwriters Labs.)

and the metal of the enclosure than did the locknuts with flat or ribbed surfaces.

The fittings investigated in this work were formed of die-cast zinc, steel and malleable iron. The melting point of zinc is 420° C while the melting point of steel and malleable iron is much higher, typically greater than 1400° C. Heat generated from the fault current in some sample assemblies was obviously greater than the melting point of the die-cast zinc fittings and locknuts, but not greater than the melting point of steel or malleable iron since no melting of the steel enclosure occurred. This was further evidenced by tests of two sample assemblies where the die cast zinc body of the fittings melted, but the steel locknuts did not.

All of the conduit fittings that were constructed of steel bodies and steel locknuts completed the test with acceptable results. A visual examination of the steel locknuts indicated that the nibs on these locknuts, which in most cases were sharp and well defined from the metal forming process, provided for better penetration through the enclosure paint than the nibs on the die-cast zinc locknuts.

For most of the sample assemblies which completed the Current Test with acceptable results, the maximum temperatures on the fitting bodies and locknuts were about the same as or less than the temperature of the conduit. In the case of the flexible metal conduit, the temperatures on the fittings were much less than on the conduit. This would seem to indicate that if the fitting can provide good electrical contact to the enclosure metal the fitting will provide for adequate equipment grounding.

Conclusions reached by Underwriters Laboratories as a result of the testing are as follows:

"1. Over 300 conduit fitting assemblies from ten different manufacturers were subjected to the Current Test to simulate performance under ground-fault conditions. As a result of the tests, only seven assemblies representing four different conduit fittings and three different manufacturers did not withstand the fault current without breaking or melting of the conduit fitting assembly. All seven of these sample assemblies were compression type connectors with die-cast zinc bodies, and all but one of these assemblies utilized a die-cast zinc locknut.

"2. An examination of the seven sample assemblies that did not complete the Current Test with acceptable results showed that the failures were probably due to high resistance from the inability of the fitting locknut to penetrate through the enclosure paint and provide good electrical continuity between the fitting and the metal enclosure. Heat generated by the high-resistance arcing was sufficient to melt the zinc, but not steel or iron.

"3. Some of the sample assemblies that did not exhibit breaking or melting did show signs of arcing and welding between the locknut and the enclosure and/or the fitting and the conduit. These sample assemblies usually had higher temperatures during the Current Test, however, the temperatures were not sufficient to cause melting of the zinc or steel parts nor loss of continuity between the conduit, fitting, and enclosure.

"4. Most of the sample assemblies that were subjected to the Current Test attained maximum temperatures on the fitting bodies and locknuts that were about the same as or less than the temperature of the conduit. For the tests with flexible metal conduit, the temperatures of the fittings were much less than the temperatures of the flexible conduit.

"5. As a result of the tests, it was observed that if the fitting provides good electrical contact to both the enclosure and the conduit, the fitting will provide a suitable equipment ground path for fault current."

Bonding for over 250 volts.

For systems having a voltage exceeding 250 volts to ground, the electrical continuity of metal raceways or metal-sheathed cables that are not service-entrance cable must also be assured. (Section 250-76) Acceptable methods include any of the methods approved for bonding at service equipment found in Sections 250-72(b) through (e). Note that standard

locknuts and bushings without additional bonding means are not permitted for bonding equipment over 250 volts to ground.

Bonding for Over-250 Volt Circuits

Listed fittings

Threadless couplings and connectors for cables with metal sheaths
Two locknuts, on rigid metal conduit or IMC, one inside and one outside
Fittings with shoulders that seat firmly against the box or enclosure such as for EMT, flexible metal conduit and cable connectors with one locknut inside

Also permitted:
• Threaded couplings and bosses
• Threadless couplings and connectors
• Bonding jumpers
• Other approved devices such as bonding-type locknuts

Figure 8-4

The bonding methods permitted include:
(b) for rigid and intermediate metal conduit, connections made up wrenchtight with threaded couplings or threaded bosses on enclosures.
(c) threadless couplings and connectors made up tight for rigid metal and intermediate metal conduit and electrical metallic tubing.
(d) bonding jumpers used around concentric or eccentric knockouts that are punched or formed so as to impair the electrical connection to ground.
(e) other approved devices like bonding type locknuts and bushings.

An exception to the requirement for bonding for these circuits provides that where oversized, concentric or eccentric knockouts are not encountered, or where concentric or eccentric knockouts have been tested and the box or enclosure is listed for the use, the following methods of ensuring continuity for these connections are permitted:
(a) threadless couplings and connectors for cables with metal sheaths.
(b) for rigid and intermediate metal conduit, two locknuts, one inside and the other outside the boxes and enclosures.
(c) fittings that seat firmly against the box or enclosure or cabinet such as for electrical metallic tubing, flexible metal conduit and

cable connectors, with one locknut inside the enclosure, or

(d) listed fittings.

Outlet boxes are available that have specially designed and tested knockouts that perform satisfactorily for over 250 volt-to-ground applications. These boxes also are required to be listed for this purpose. It is required that listed fittings that are used to comply with the exception to the rule for bonding comply with the requirements of suitability found in Section 110-3.

Oversized knockouts.

The installer needs to be cautious in the use of equipment that has concentric or eccentric knockouts as their ability to carry fault current must be of concern. It is very common to find nibs of adjacent rings damaged during removal of the desired knockout. This leaves less material available for carrying fault current. The safest practice is to install bonding bushings around concentric and eccentric knockouts where there is any question about their integrity.

Concentric and Eccentric Rings

Concentric rings. May be necessary to bond around rings to ensure current-carrying capacity.

Eccentric rings present the same obstacle to carrying fault current as concentric rings.

Figure 8-5

In other areas, where oversized, concentric or eccentric knockouts are not present, threadless fittings which are made up tight with conduit or armored cable or the use of two locknuts, one inside and one outside of boxes and cabinets, are acceptable for bonding.

Concentric and eccentric knockouts in equipment such as panelboards, auxiliary gutters and wireways are not tested or certified by an electrical products testing laboratory for their current-carrying ability. As such, they should not be relied upon for such a purpose and should have bonding bushings and bonding jumpers installed around them.

Where loosely-jointed metal raceways are used and especially where there may be expansion joints or telescoping sections of raceways, the Code requires that they be made electrically continuous by the use of equipment bonding jumpers or other means. (Section 250-77)

Reducing washers.

Reducing washers are commonly used in electrical installations where it is desirable or necessary to install conduit or fittings of a size that is smaller than the knockout available in the enclosure. These reducing washers are not suitable for bonding and must always be bonded around to provide an adequate fault-return path.

Fault currents.

Those equipment bonding jumpers form a part of the equipment grounding conductor and are, therefore, called upon to carry the same fault current that the equipment grounding conductor would carry and must be the same size.

The size of the bonding jumper will depend on its location and is always based on the size of the nearest overcurrent device in the circuit ahead of the equipment. [Section 250-79(e)] Column 1 of Table 250-95 gives the size or setting of the overcurrent device in the circuit ahead of the equipment. Columns 2 and 3 give the minimum size of the equipment grounding conductor whether copper or aluminum.

Attaching jumpers.

Where bonding jumpers are used between grounding electrodes or around water meters and the like, the Code requires that good electrical contact be maintained and that the arrangement of conductors be such that the disconnection or removal of equipment will not interfere with or interrupt the grounding continuity of the jumper. (Section 250-112)

Bonding jumpers shall be attached to circuits and equipment by means of exothermic welding, listed pressure connectors, listed clamps or other suitable and listed means. A connection that depends solely on solder is not acceptable. Sheet metal screws are not permitted to connect grounding conductors to enclosures. (Section 250-113)

Bonding multiple raceway systems.

Where more than one raceway enters or leaves a switchboard, pull or junction box or other equipment, it is permissible to use a single conductor to bond these raceways to the equipment. The equipment grounding conductor is sized for the largest overcurrent device ahead of conductors contained in the raceways. See Section 250-79(e).

Bonding Multiple Raceways

Figure 8-6

For example, as shown in Figure 8-6, four metallic raceways leave the bottom of an open switchboard or motor control center. The overcurrent protective devices ahead of the raceways are 400, 300, 225 and 125 amperes respectfully. According to Table 250-95, the minimum size equipment bonding jumper for the raceway having conductors protected at 400 amperes is No. 3 copper or No. 1 aluminum. If this conductor were looped through a grounding bushing on each raceway, compliance with the Code would be obtained. Of course, the grounding bushings would need to be listed for both the size of conduit and conductor.

In addition, as shown in Figure 8-7, it is acceptable to install an equipment bonding jumper individually from each raceway to the equipment grounding terminals of the equipment. Each bonding conductor is sized for the overcurrent device ahead of the conductors in the raceway per Table

Individual Bonding Jumper

Overcurrent Protection	Equipment Bonding Jumper
400 Amperes	No. 3 copper - No. 1 aluminum
300 Amperes	No. 4 copper - No. 2 aluminum
225 Amperes	No. 4 copper - No. 2 aluminum
125 Amperes	No. 6 copper - No. 4 aluminum

Figure 8-7

250-95. In some cases, this method may require a larger conductor to bond some conduits than where individual bonding jumpers are installed.

Nonmetallic enclosures.

Section 370-3 permits metal raceways or metal-jacketed cables to be used with nonmetallic enclosures only where:

1. internal bonding means are provided between all raceways or cables.
2. internal bonding means with provision for attaching a grounding jumper inside the box are provided.

The bonding jumpers must, of course, be sized in accordance with Table 250-95. It should be noted that the size of the bonding jumpers given in Section 250-95 is the minimum size. Larger bonding jumpers may be required to comply with the available fault-current requirements in Section 110-10.

Bonding receptacles.

An equipment bonding jumper is required to connect the grounding terminal of a grounding type receptacle to a grounded box. (Section 250-74) Where more than one equipment grounding conductor enters a box, they must be spliced or joined inside the box with suitable devices. Four exceptions are provided.

Exception No. 1. Where the box is surface mounted and direct metal-to-metal contact is made

Bonding of Grounding Receptacles

- Where more than one conductor enters the box, they must be spliced together using suitable devices. A bonding jumper is required to be connected to the grounding terminal of grounding receptacles.

- Ex. 1. Surface-mounted boxes and direct metal-to-metal contact is achieved. Cover-mounted receptacles not approved unless listed.

- Ex. 2. Receptacles that have contact devices or yokes listed as providing grounding.

- Ex. 3. Floor boxes that are listed as providing satisfactory grounding.

- Ex. 4. Receptacles that have an isolated equipment grounding terminal for reduction of electromagnetic interference.

Figure 8-8

between the receptacle and the box. (Cover mounted receptacles are not approved unless the box and cover combination are listed as providing adequate ground continuity.)

Exception No. 2. Contact devices or yokes designed and listed as providing bonding with the mounting screws to establish the grounding circuit between the device yoke and flush-type boxes.

Exception No. 3. Floor boxes that are listed as providing satisfactory grounding continuity.

Exception No. 4. A receptacle having an isolated equipment grounding terminal for reduction of electromagnetic interference. In this case, the equipment grounding terminal must be grounded by an insulated equipment grounding conductor.

Self-Grounding Type Receptacle

Figure 8-9. *(Courtesy Eagle Electric Co.)*

This equipment grounding conductor is permitted to pass through one or more panelboards to terminate at an equipment grounding terminal of the separately derived system or service. However, this insulated equipment grounding conductor must be connected to the equipment grounding terminal bar at the building disconnecting means or the source of a separately derived system within the same building or structure.

To provide effective grounding, the insulated equipment grounding conductor must never pass without being terminated at the separately derived system that is the source of power for the equipment being grounded.

Isolated Ground Receptacle

Figure 8-10. *(Courtesy Pass & Seymour/ Legrand)*

Installation of equipment bonding jumper.

The equipment bonding jumper is permitted to be installed either inside or outside of a raceway or enclosure. [Section 250-79(f)]

Where the jumper is installed on the outside, the length is limited to not more than 6 feet. In addition, the bonding jumper must be routed with the raceway. This is vital to keep the impedance of the equipment bonding jumper as low as possible. See Figure 8-11.

Bonding of piping systems.

Section 250-80 requires that interior metal water piping and other metal piping systems be bonded.

Installation of Bonding Jumper

Where installed outside the raceway or enclosure, the length is limited to not more than 6 feet and the bonding jumper must be routed with the raceway.

Figure 8-11

This requirement for bonding is not to be confused with the requirement in Section 250-81 that metal underground water piping is to be used as a grounding electrode. Some requirements change depending upon whether the piping is interior metal water piping or other metal piping systems.

Bonding Metal Piping Systems

Figure 8-12

Included among the items within a building that should be adequately grounded to the one common grounding electrode are the interior water piping system (hot and cold) and gas and sewer piping, any metallic air ducts within the building, as well as such devices as TV towers, gutters provided with a deicing system, etc.

Bonding Metal Water Pipes

Connections accessible. Size per Section 250-94.

Figure 8-13

Metal water piping.

This requirement for bonding applies to all interior metal water piping systems. [Section 250-80(a)] The bonding jumper is to be sized in accordance with Table 250-94 and, thus, is based on the size of the service-entrance conductor and not on the rating of the service overcurrent device. In addition, the points of attachment of the bonding jumper to the metal water piping system are required to be accessible. The piping system is permitted to be bonded to the service equipment enclosure, the grounded conductor at the service, the grounding electrode conductor where large enough, or to one or more grounding electrodes.

For example, if Table 250-94 requires a No. 2 bonding jumper to the metal water piping, it cannot be connected to a No. 6 or No. 4 grounding electrode conductor.

Where a metallic underground water piping system exists and is connected to a metallic interior water-piping system which does not contain an insulated coupling, the interior water piping system was automatically and adequately grounded when the metallic underground water piping system was used as the grounding electrode. However, with the expanding use of nonmetallic piping and insulated couplings it becomes more important to be sure that the interior piping not only is electrically continuous, but that also it is adequately grounded by bonding it to the same grounding electrode used for the premises. That is a mandatory and essential requirement of the Code.

Bonding metal water piping from separately derived system.

Where a separately derived system uses a made

Bonding Metal Water Piping From Separately Derived System

Bond interior metal water pipe at the nearest available point in the area served by the separately derived system, to the grounded conductor -

where the system is not grounded to an effectively grounded structural metal member or effectively grounded metal water pipe.

Figure 8-14

grounding electrode as covered in Section 250-26(c)(3), the nearest available point on the interior metal water piping system in the area served by the separately derived system shall be bonded to the grounded conductor of the separately derived system. This rule applies where there is no effectively grounded metal water pipe or effectively grounded building steel in the vicinity of the separately derived system. In this case, a "made electrode," such as a ground rod, is being used to ground the separately derived system as permitted by Section 250-26(c)(3).

The bonding conductor from the separately derived system to the piping shall be sized in accordance with Table 250-94 for the derived phase conductors. It must be installed in accordance with the general rules for installing grounding conductors in Sections 250-92(a) and (b). The points of attachment of the bonding conductor shall be accessible.

This requirement assures that metal piping, such as process or manufacturing equipment piping, in the area served by the separately derived system is bonded.

Multiple occupancy building.

An exception to Section 250-80(a) allows the metal interior water piping system to be bonded to the panelboard or switchboard enclosure (other than service equipment) under specific conditions. The conditions are as follows:

Bonding Metal Interior Piping in Multiple-Occupancy Building

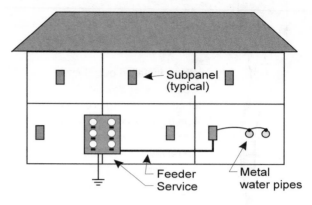

Figure 8-15

1. The building is multiple occupancy, and
2. The metallic water piping is isolated from all other occupancies by nonmetallic water piping.

In this case, the bonding jumper to the water piping is sized in accordance with Table 250-95. Thus, the ampere rating of the overcurrent device supplying the feeder to the unit or occupancy determines the minimum size of the bonding conductor. The bonding jumper for the interior metal water piping runs from the equipment grounding terminal bar in the panelboard serving the unit to the piping. In this case, the bonding jumper does not connect to the neutral terminal bar in the panelboard.

Other metal piping.

Other interior metal piping which may become energized must be bonded. The Code does not give guidance on how to determine the conditions under which metal piping is likely to become energized. Since metal piping systems are conductive, from a safety standpoint, all metal piping systems should be bonded.

Common systems that must be bonded include interior metal:

    Gas piping systems
    Pneumatic systems
    Waste, drain and vent lines
    Oxygen, air and vacuum systems.

The bonding conductor is sized from Table 250-95 using the rating of the overcurrent device in the circuit ahead of the equipment. The equipment grounding conductor for the circuit that may energize the equipment can be used as the bonding conductor. The point of connection of these bonding conductors to the metal piping systems is not

Bonding Other Metal Piping

Size bonding jumper per Section 250-95.

Figure 8-16

Bonding Structural Steel

- Exposed interior structural steel that is interconnected to form a steel building frame and is not intentionally grounded and may become energized shall be bonded.
- Bond to the service equipment enclosure, the grounded conductor at the service, the grounding electrode conductor where of sufficient size, or to the one or more grounding electrodes used.
- Size bonding jumper per Table 250-94.
- Install per Section 250-92(a) and (b).
- Attachment point(s) of bonding jumper to be accessible.

Figure 8-17

required to be accessible as the connections to metal water piping systems are.

The bonding conductor is permitted to be connected to the:

1. service equipment enclosure.
2. grounded conductor at the service.
3. grounding electrode conductor, where of adequate size.
4. one or more grounding electrodes.

Bonding structural steel.

Exposed interior structural steel that is interconnected to form a steel building frame and is not intentionally grounded and may become energized is required to be bonded to the service equipment enclosure, the grounded conductor at the service, the grounding electrode conductor where of sufficient size, or to the one or more grounding electrodes used.

The bonding jumper must be sized in accordance with Table 250-94 and installed in accordance with the rules in Section 250-92(a) and (b). The points of attachment of the bonding jumper to the structural steel shall be accessible.

Chapter Eight: The questions included here were developed using material included in this chapter. The answers can be found by reviewing the text. It is also important that students make use of the 1996 NEC®, where many answers can be found. See page 236 for answers.

1. For systems having a voltage exceeding ____ volts to ground, the electrical continuity of metal raceways or metal-sheathed cables, that are not service-entrance cable, must also be assured.
   a. 125
   b. 150
   c. 100
   d. 250

2. Bonding jumpers are required to be attached to circuits and equipment by means of ____, or other listed means.
   a. exothermic welding
   b. listed pressure connectors
   c. listed clamps
   d. any of the above

3. Metal raceways or metal-jacketed cables are permitted to be used with nonmetallic enclosures only where ____.
   a. internal bonding means are provided between all raceways
   b. internal bonding means with provision for attaching a grounding jumper inside the box are provided
   c. internal bonding means are provided between all cables
   d. any of the above

4. Without exception, where more than one equipment grounding conductor enters a box, they must be spliced or joined with____ devices.
   a. labeled
   b. listed
   c. suitable
   d. bonding

5. An equipment grounding conductor is permitted to pass through one or more panelboards within the same ____ to terminate at an equipment grounding terminal of the separately derived system or service.
   a. equipment
   b. building
   c. enclosure
   d. cabinet

6. Where the equipment bonding jumper is installed on the outside of a raceway, the length is limited to not more than ____ feet. In addition, the bonding jumper must be routed with the raceway.
   a. 7
   b. 8
   c. 6
   d. 10

7. The metal water piping system is permitted to be bonded to the service equipment enclosure, the grounded conductor at the service, the grounding electrode conductor where large enough, or to one or more grounding ____.
   a. clamps
   b. devices
   c. fittings
   d. electrodes

8. Common systems that must be bonded include interior metal ____.
   a. gas piping and pneumatic systems
   b. waste, drain and vent lines
   c. oxygen, air and vacuum systems
   d. all of the above

9. The equipment grounding conductor for the circuit that may energize the equipment can be used as the ____ conductor.
   a. equipment bonding
   b. grounded
   c. bonding
   d. identified

10. Expansion joints or telescoping sections of metal raceways must be made electrically continuous by the use of a ____.
   a. steel strap
   b. equipment bonding jumper
   c. welding cable
   d. equipment grounding conductor

11. The locknut/bushing and double-locknut types of installations are not acceptable for bonding in ____.
   a. hazardous (classified) locations
   b. commercial locations
   c. industrial location
   d. computer rooms

12. If four metal raceways have their conductors protected by overcurrent protective devices sized at 400, 300, 225 and 125 amperes and leave the bottom of an open switchboard or motor control center, the minimum size of a single equipment bonding jumper to be used to bond them together is ____.
   a. No. 4 copper or No. 4 aluminum
   b. No. 6 copper-clad aluminum
   c. No. 3 copper or No. 1 aluminum
   d. No. 4 copper or No. 3 aluminum

13. The points of attachment of the bonding jumper to the interior metal water piping system are required to be ____.
   a. acceptable
   b. marked
   c. accessible
   d. soldered

14. Concentric and eccentric knockouts in enclosures like wireways and panelboards:
   a. are always suitable for grounding and bonding.
   b. are tested by a qualified electrical testing laboratory for their current-carrying ability.
   c. are not tested by a qualified electrical testing laboratory for their current-carrying ability.
   d. are capable of carrying large amounts of fault current.

15. Bonding connections to interior metal water piping systems:
   a. are permitted to be made with solder.
   b. must be accessible.
   c. are permitted to be connected to the neutral in a subpanel.
   d. are not required.

16. Structural steel must be bonded where:
   a. it is exposed on the interior of the building.
   b. it is interconnected to form a steel building frame.
   c. it is not intentionally grounded.
   d. all of the above.

Chapter **9**

Equipment Grounding Conductors

Objectives

After studying this chapter, the reader will be able to understand

- General requirements for equipment grounding conductors on grounded and ungrounded systems.
- Rules applied to multiple raceways or cables.
- Rules for flexible cords.
- Use of building steel that is properly grounded by an equipment grounding conductor.
- Grounding of equipment by the grounded circuit conductor.

Definition.

"**Grounding conductor, equipment.** The conductor used to connect the noncurrent-carrying metal parts of equipment, raceways, and other enclosures to the system grounded conductor, the grounding electrode conductor, or both at the service equipment or at the source of a separately derived system."*

Equipment Grounding Conductor Path

Must be:
- **permanent and continuous,**
- **have ample capacity to conduct safely any currents likely to be imposed on it,**
- **of the lowest practical impedance.**

General requirements.

Equipment grounding conductors are intended to prevent an objectionable potential above ground on conductor and equipment enclosures and to provide a low-impedance path for fault-currents by connecting the equipment grounding conductor to the grounded conductor if it is a grounded system. The equipment grounding conductor or path must also:

1. be permanent and electrically continuous;
2. have ample capacity to conduct safely any currents liable to be imposed on it; and
3. be of the lowest practical impedance (Section 250-51).

The equipment grounding conductor or path must thus connect with the grounding electrode in a low-impedance path, and if the system is grounded, it must also be connected through a low-impedance path to the grounded service conductor.

Grounded system.

The equipment grounding conductor or path must extend from the furthermost point on the circuit to the service equipment where it is connected to the grounded conductor. This connection

---

is made through the main bonding jumper. Often, this equipment grounding conductor or path is the conductor enclosure (conduit, cable jacket, etc.).

Ground Fault in Grounded System

Figure 9-1

Should insulation failure occur anywhere on a phase conductor and a ground fault develop between the energized conductor and the conductor enclosure, a ground-fault circuit will be established.

The ground-fault circuit will thus be from the source, through the supply conductors, through the overcurrent devices, to the point of fault on the phase conductor, usually through an arc, through the equipment grounding conductor, through the main bonding jumper, to the grounded conductor or neutral and back to the source.

If this circuit is complete, of adequate capacity and low impedance, the equipment and persons who may contact it are protected. A break in this equipment grounding circuit or other grounding system failure will cause equipment failures and expose persons to possibly lethal shocks.

Ungrounded systems.

In an ungrounded system, the equipment grounding conductor or path must be permanent and continuous to ground in order to keep all equipment and conductor enclosures at ground potential. The equipment grounding conductor must be the same size as called for in a grounded system if we are to have maximum safety. The sizes of equipment grounding conductors given in Table 250-95 apply equally to an ungrounded system and a grounded system.

In general, it may be said that any conductor or equipment enclosure, be it conduit, electrical metallic tubing raceway or busduct enclosure, provides a satisfactory equipment grounding conductor for an ungrounded system if all joints are made

Ground Fault in Ungrounded System

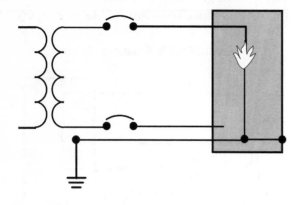

Figure 9-2

electrically continuous. It may be necessary to use bonding jumpers at certain points. Such bonding jumpers must be sized per Table 250-95.

Equipment grounding conductor material.

The Code specifies in Section 250-91(b) the conductors which shall be permitted to be used for equipment grounding conductors. They are as follows:

1. a conductor of copper or other corrosion-resistant material such as aluminum that is solid, stranded, insulated, covered or bare (some sections of the Code may specifically require an equipment grounding conductor to be insulated or solid) in the form of a wire or busbar of any type;
2. rigid metal conduit;
3. intermediate metal conduit;
4. electrical metallic tubing;
5. flexible metal conduit where both the conduit and fittings are listed for grounding;
6. armor of Type AC cable;
7. the copper sheath of Type MI Cable;
8. type MC metal sheath or the combined metal sheath and grounding conductor;
9. cable trays as permitted in Article 318;
10. cablebus frameworks as permitted in Section 365-2(a); and
11. other electrically continuous metal raceways listed for grounding.

It should be noted here that these requirements or provisions in Section 250-91(b) are general in nature. Many sections of the Code contain specific requirements that must be complied with. For example, Section 517-13(a) requires a supplementary insulated equipment grounding conductor in patient care areas, as does Section 550-24(a) for

the feeder to a mobile home unless the service is mounted on the mobile home as permitted in Exception No. 2 to Section 550-23.

Flexible Metal Conduit as Equipment Ground

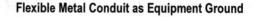

- Flexible metal conduit must be listed.
- Fittings listed for grounding.
- Combined length of flexible metal conduit, flexible metallic tubing and liquidtight flexible metal conduit in same ground return path not more than 6 feet.
- Maximum 20 ampere overcurrent protection of contained conductors.
- For connection of equipment where flexibility is required, additional equipment grounding conductor required.

Figure 9-3

As permitted in Section 250-91(b) Exception No. 1, listed flexible metal conduit, listed flexible metallic tubing and listed liquidtight flexible metal conduit in the 3/8 through 1¼-inch trade sizes may be used as the equipment grounding conductor where:

1. the combined length of the ground return path for these wiring methods does not exceed 6 feet,
2. the conduit or tubing is terminated in fittings listed for grounding, and
3. the circuit conductors contained in these wiring methods are protected by an overcurrent device

Liquidtight Flexible Metal Conduit

- Liquidtight flexible metal conduit must be listed.
- Fittings listed for grounding.
- Maximum overcurrent protection of contained conductors:
  - 20 amperes for 3/8 and ½ inch
  - 60 ampere for ¾ thru 1¼ inch
- Maximum 6 feet of LFMC in any ground return path.
- For connection of equipment where flexibility is required, additional equipment grounding conductor required.

Figure 9-4

rated at 20 amperes or less. See Figure 9-5.

In Section 250-91(b) Exception No. 2, listed liquidtight flexible metal conduit in the ¾-inch through 1¼-inch trade sizes may be used for equipment grounding where:

1. the total ground return path does not exceed 6 feet,
2. no flexible metal conduit or flexible metallic tubing, or liquidtight flexible metal conduit in the 3/8-inch and ½-inch trade sizes are used in this ground return path, and
3. fittings that are listed for grounding are used for connections.

The ¾-, 1- and 1¼-inch sizes are permitted to be used as an equipment grounding conductor only where the contained conductors have overcurrent protection not more than 60 amperes. Liquidtight flexible metal conduit in the 1½-inch and larger trade sizes must always have an equipment grounding conductor installed for the equipment that requires grounding.

Conductor enclosures.

Conduit runs, which are properly threaded and in which the couplings are set up tightly, preferably using a joint sealer that will not reduce continuity, may be expected to perform satisfactorily as an equipment grounding conductor for runs of limited length. A joint sealer aids in assuring an effective equipment grounding path because it acts as a lubricant and permits the joint to be screwed up tighter. Under poor conditions, the conduit impedance with couplings should not show an increase of over 50 percent when compared with a straight run of conduit. The use of this higher impedance figure would provide a factor of safety. In the case cited, there is no economic justification for using an additional equipment grounding conductor.

The Code further requires that where conduit is used as a grounding conductor, all joints and fittings shall be made wrenchtight. [Section 250-92(c)(1)] This calls attention to the fact that conduit, where used as an equipment grounding means, is a current-carrying conductor under fault conditions and must be made electrically continuous by having joints made up tight.

Usually, large and often parallel conduits are installed from the transformer to the service equipment. Then, smaller and smaller conduits are installed for feeders and branch circuits. For instance, at one point the equipment grounding path

Equipment Grounding Conductor Path

Figure 9-5

may be three 4-inch conduits in parallel; at another point, two 4-inch conduits in parallel while down the line it may be only one 1¼-inch conduit, all being connected together to form a permanent and continuous path. As the circuit changes from large overcurrent protective means to smaller ones, the conductivity of the equipment grounding path becomes smaller and smaller. The conduit or tubing at the end of the circuit may be no larger than ½-inch electrical metallic tubing or 3/8-inch flexible metal conduit.

The NEC® does not dictate any particular size of conduit to serve as the equipment grounding conductor for an upstream overcurrent device, other than as mentioned in the previous section of this text. It is felt that the metallic raceway that is sized properly for the conductor fill will provide an adequate equipment ground fault return path.

Size of equipment grounding conductor.

The entire equipment grounding conductor or path of any raceway system will be as shown diagrammatically in Figure 9-6. Starting at the service we have a large overcurrent protective device which is in series with other and usually smaller feeder or branch overcurrent protection devices. The ungrounded (hot) conductor usually decreases in size as it progresses through smaller and smaller overcurrent devices.

Thus, a ground fault may be cleared through a 15 ampere overcurrent device so that the equipment ground path below the 15 ampere overcurrent device only needs to be a No. 18 copper wire (subject to Code limitations) as given in Section 250-95,

Minimum Size of Equipment Grounding Conductor

Minimum size of equipment grounding conductor based on rating or setting of overcurrent device in circuit ahead of equipment.

Figure 9-6

Exception No. 1. That conductor has a short-time rating of about 38 amperes and it should thus be adequate to operate the 15-ampere overcurrent device if it operates in a reasonable time.

Table 250-95 provides the minimum size of conductor that is permitted to be used as an equipment grounding conductor. The size is based on the ampere rating of the overcurrent protective device ahead of the conductor. Table 250-95 is reprinted as Table 6 in Chapter 18 of this text.

For example, if the overcurrent protection ahead of the circuit or feeder is 225 amperes, the minimum size equipment grounding conductor is found as follows:

In Table 250-95, follow the first column, which gives the rating of the overcurrent device, down to find the rating which equals or exceeds 225 amperes. Since 225 amperes is not found, go to the next larger size which is 300 amperes. Follow that line across to find the minimum size copper wire to be No. 4 and for aluminum, a No. 2 wire.

Follow a similar process to determine the minimum size conductor for any installation. A comprehensive analysis of the withstand rating of these equipment grounding conductors can be found in Chapter 11 of this text.

Three exceptions to the general sizing rule for equipment grounding conductors are found in Section 250-95. They are as follows:

1. An equipment grounding conductor not smaller than No. 18 copper and not smaller than the circuit conductors is permitted for fixture wires or cords in accordance with Section 240-4.

2. The equipment grounding conductor is not required to be larger than the circuit conductors supplying the equipment.

3. Where a raceway or a cable armor or sheath is used as the equipment grounding conductor as provided in Sections 250-51, 250-57(a) and 250-91(b).

Adjusting size of equipment grounding conductor.

Section 250-95 requires that, "Where conductors are adjusted in size to compensate for voltage drop, equipment grounding conductors, where required, shall be adjusted proportionately according to the circular mil area." This means that where a feeder or branch-circuit conductor is increased in size to compensate for voltage drop, the equipment grounding conductor, where run, must be increased at not less than the same ratio the feeder or circuit conductors are increased.

For example, a 200 ampere feeder is to be installed. It is determined that the voltage drop would be excessive. A 250 kcmil conductor is selected for the feeder rather than installing the 3/0 copper conductor as is permitted by Table 310-16. Table 250-95 requires a No. 6 equipment grounding conductor for the 200 ampere overcurrent device.

Determine the minimum size equipment grounding conductor required for the feeder by the following formula: (Use Table 8 of Chapter 9 to determine the area in circular mils where the conductor size is given by a non-circular mil designation.)

$$\frac{\text{Selected Feeder Conductor Area}}{\text{Required Feeder Conductor Area}} = \text{Ratio}$$

Table 250-95 EGC x Ratio = Required EGC

$$\frac{250000 \text{ kcmil}}{167800 \text{ kcmil}} = 1.49 \text{ percent}$$

.027 (No. 6) x 1.49 = .04023 circular mills
Next larger size = No. 3 copper required.

Supplemental grounding electrode.

Engineers often specify that ground rods be installed to ground metal lighting standards or poles and at metal poles for electric signs. These rods are permitted to be used but must be considered to be supplemental ground rods. They can "supplement" the equipment grounding conductor

Supplemental Grounding Electrode

Supplemental grounding electrode permitted

Equipment grounding conductor in wiring method with circuit conductors

Figure 9-7

that is run with the branch circuit but cannot be the only means of grounding this or similar equipment. See Section 250-91(c). To use these ground rods as the only means of grounding would constitute an earth return which is unsafe and prohibited by Code.

Equipment grounding conductor with circuit conductors.

A very important requirement for installing equipment grounding conductors is contained in Section 250-57(b). The requirement is that the equipment grounding conductor must generally be installed in the same raceway, cable or cord, or otherwise be run with the circuit conductors. This requirement is repeated in Section 300-3(b) where, in addition to the requirement for raceways, equipment grounding conductors are required to be contained in the same trench with other circuit conductors. This requirement is critical for the installation of alternating current systems.

It has been proven that separating the equipment grounding conductor from the circuit conductors greatly increases the impedance of the circuit. This excessive separation can render an adequately-sized equipment grounding conductor ineffective in carrying enough current to operate the circuit protective device and clear the faulted equipment. In this case, providing a properly-sized equipment grounding conductor, but installing it improperly, results in an ineffective and possibly unsafe installation.

See Chapter 11 of this text for additional information on this subject.

Nonmetallic raceway.

Where the wiring system is nonmetallic, it is necessary to run an equipment grounding conductor along with the circuit conductors. Do not separate them at any point in the circuit by any metallic material regardless of whether the metallic material is magnetic or not. It is true that if the material is nonmagnetic, the increase in impedance of the circuit will not be as great as if the material was magnetic. In any case such separation must be avoided.

Multiple raceways or cables.

Special rules apply where more than one raceway or cable is installed with parallel conductors (two or more conductors that comply with Section 310-4 and are connected together at each end to form a single conductor) and an equipment grounding conductor is installed in the raceway. In this case, Section 250-95 requires that an equipment grounding conductor be installed in each raceway or cable. Each equipment grounding conductor must be sized in compliance with the ampere rating of the overcurrent device protecting the conductors in the raceway or cable.

Section 310-4 permits equipment grounding conductors, smaller than 1/0, to be sized in compliance with Table 250-95. However, all other requirements for installing conductors in parallel must be met. These rules require that each set: (1) be the same length; (2) be of the same conductor material

Size of Equipment Grounding Conductors

600 ampere OC device rating

No. 1 CU or No. 2/0 AL equipment grounding conductor in each raceway per Table 250-95

Raceways with parallel equipment grounding conductors

No. 4 CU or No. 2 AL equipment grounding conductor in cable by manufacturer

No. 1 CU or No. 2/0 AL required if installed in parallel

Cables with parallel equipment grounding conductors

Equipment (typical)

Figure 9-8

(all copper or all aluminum); (3) be the same size in circular mil area; (4) have the same insulation type; (5) be terminated in the same manner; and (6) the raceways or cables must have the same physical properties.

One reason for this requirement for installing equipment grounding conductors in parallel is shown in Figure 9-8. In the event of a line-to-ground fault in the equipment supplied by the circuit, the fault current will divide equally between the equipment grounding conductors. However, should a line-to-ground fault occur in the raceway or cable, current will be fed to the fault from both directions. The equipment grounding conductor will thus be called upon to carry the entire amount of fault current until the overcurrent protective device ahead of the fault opens.

In some cases, where cables are installed in parallel, special constructions will be required to get the larger equipment grounding conductor in the cable. Listed cables are generally produced with the equipment grounding conductors sized in compliance with a construction standard that complies with Table 250-95. For example, a copper cable construction suitable for a 300 ampere overcurrent device will have a No. 4 copper equipment grounding conductor within the cable by the manufacturer. If two of these cables are installed in parallel and connected to a 600 ampere overcurrent protective device, a No. 1 copper equipment grounding conductor would be required in each cable to comply with Table 250-95.

Flexible cord.

The use of an equipment grounding conductor in a supply cord is permitted providing the cord is connected as specified in Section 400-7. The method of grounding noncurrent-carrying metal parts of portable equipment may be by means of the equipment grounding conductors in the flexible cord feeding such equipment. The proper type attachment plug must be used to terminate the conductors and the attachment plug also have provision to make contact with a grounding terminal in the receptacle.

For the grounding of portable or pendant equipment, the conductors of which are protected by fuses or circuit breakers rated or set at not exceeding 20 amperes, the Code permits the use of a No. 18 copper wire as an equipment grounding circuit conductor, provided the No. 18 grounding conductor is a part of a listed flexible cord assembly. If such portable or pendant equipment is served by conductors protected at more than 20 amperes, then the equipment grounding circuit conductor used must conform to the sizes given in Table 250-95.

See Section 250-95 Exception No. 2 where the equipment grounding conductor is not required to be larger than the circuit conductors.

Use of building steel.

Section 250-58 permits a metal rack or structure to ground electric equipment that is secured to it and in electrical contact, provided the support means is grounded by an equipment grounding conductor as specified by Section 250-57. However, the building steel is not permitted to serve as an equipment grounding conductor to ground equipment. That is due to the uncertain path that ground fault current must take in an effort to clear a fault.

Figure 9-9

This section emphasizes the requirement in Sections 250-57(b) and 300-3(b) that the equipment grounding conductor must be in the same raceway, cable or cord, or otherwise be run with the circuit conductors. Again, this is so the grounding circuit impedance will be as low as possible to allow adequate ground-fault current to flow so the circuit protective device will clear the fault.

In the same manner, and on the same basis, metal car frames, supported by metal hoisting cables attached to or running over sheaves or drums of elevator machines, are considered grounded when the machine is grounded as required by the Code.

Grounding for direct-current circuits.

All of the above applies to the grounding of alternating-current systems where reactance of the circuit plays a large part. In the case of direct-current circuits we are concerned with ohmic resistance only. Owing to that fact, the current-carrying capacity of the grounding conductor for a direct-current supply system must be equal to that of the largest conductor of the system. However, if the grounded circuit conductor is a neutral derived from a balancer winding or a balancer set which has overcurrent protection as required in the Code under Section 445-4(d), then the grounding conductor size shall be not less than the size of the neutral. The requirement for overcurrent protective devices here states that the two-wire direct-current generators used in conjunction with balancer sets shall be equipped with overcurrent protective devices which will disconnect the 3-wire system in the case of excessive unbalancing of voltages or currents.

Grounding conductors for direct-current circuits.

For direct-current circuits, the size of the grounding conductor is specified in Section 250-93. The size may be larger than would be required for the same size alternating-current circuit. That is because resistance is the only factor in determining current flow in a direct-current circuit. Therefore, fault-currents are larger in a dc circuit. In an alternating-current circuit, impedance becomes the important factor for not only resistance, but reactance also must be taken into account.

Long term reliability of metal raceways.

In the above discussions, it is assumed that a conductor enclosure (conduit or other raceway) has been properly installed with good tight joints which will provide a permanent and continuous electrical circuit when it is first installed. However, time will take its toll and will tend to destroy the continuity that was begun with.

The safety of an electrical system will thus depend on how long we can expect the equipment grounding conductor to remain permanent and continuous. The answer will vary depending on the type of metal raceway, the environment it is installed in and the quality of the installation.

For design purposes we can break the types into two categories:

1. Where little corrosion will exist and where we can reasonably expect that the equipment grounding conductor, in the form of a metal raceway, remain permanent and continuous for a period of fifty years or more.

2. Where corrosion in varying degrees will exist and where the permanency of the equipment grounding conductor provided by the metal raceway can be questioned.

In the first category, we may place most commercial and residential buildings. That being the case, we could use all conductor enclosures, which are approved for the purpose, as part of the equipment grounding conductor (with the use of bonding jumpers where required).

In the second category, we could include all industrial plants where a metallic equipment grounding conductor, sized per Table 250-95, must be run in parallel with and within the conductor enclosure so as to ensure continuity if that circuit is broken owing to corrosion sooner or later.

Metal conduit underground.

Care must be taken when installing metallic conduit and electrical metallic tubing in the earth, in concrete on or below grade, or where exposed to moisture. [Section 346-1(c)] The Underwriters Laboratories' 1995 guide card information for rigid metal conduit (DYIX) and intermediate metal conduit (DYBY) contains the following information regarding corrosion protection:

"Galvanized rigid (and intermediate) steel conduit installed in concrete does not require supple-

Metal Conduit Corrosion Protection Required		
In Concrete:	**Required**	**Optional**
Rigid Steel		X
Intermediate Steel		X
Aluminum Rigid	X	
Steel EMT	Below grade may need	On or above grade
Alum. EMT	X	
In Soil:		
Rigid Steel		X
Intermediate Steel		X
Aluminum Rigid	X	
Steel EMT	Severely corrosive soil	
Alum. EMT	X	

mentary corrosion protection. Galvanized rigid (and intermediate) steel conduit installed in contact with soil does not generally require supplementary corrosion protection.

In the absence of specific local experience, soils producing severe corrosive effects are generally characterized by low resistivity less than 2000 ohm-centimeters.

Wherever ferrous metal conduit runs directly from concrete encasement to soil burial, severe corrosive effects are likely to occur on the metal in contact with the soil.

Supplementary nonmetallic coatings as part of the conduit have not been investigated for resistance to corrosion."

In addition, experience has shown that steel conduit fails rapidly where exposed to corrosive environments found at some seacoast marinas, boatyards and plants as well as at some chemical plants. In addition, experience has shown that metal conduit systems are particularly vulnerable to failure from corrosion where they pass from concrete to exposure to an atmosphere containing corrosive elements, particularly in combination with atmospheres containing oxygen.

For electrical metallic tubing, (See Section 348-1) the following instructions are given in the UL 1995 guide card (FJMX):

"Galvanized steel electrical metallic tubing installed in concrete, on grade or above, generally requires no supplementary corrosion protection. Galvanized steel electrical metallic tubing in concrete slab below grade level may require supplementary corrosion protection.

In general, galvanized steel EMT in contact with soil requires supplementary corrosion protection. Where galvanized steel electrical metallic tubing without supplementary corrosion protection extends directly from concrete encasement to soil burial, severe corrosive effects are likely to occur on the metal in contact with the soil.

Aluminum electrical metallic tubing used in concrete or in contact with soil requires supplementary corrosion protection.

Supplementary nonmetallic coatings presently used have not been investigated for resistance to corrosion."

As a result, the Authority Having Jurisdiction must make a decision regarding the suitability of these raceways for these applications. This, of course, affects the reliability of the raceway serving as an equipment grounding conductor. Several

reports have been made where electrical metallic tubing installed to provide an equipment grounding means has failed due to corrosion. To maintain the integrity of the equipment grounding means, some inspection agencies require that a copper equipment grounding conductor be installed in parallel with the electrical metallic tubing.

Also, the authority having jurisdiction must make a decision regarding the suitability of supplementary nonmetallic coatings intended for resistance to corrosion.

Grounding of equipment by the grounded circuit conductor.

The Code does not generally permit the grounding of the grounded circuit conductor on the load side of the service disconnecting means. [Sections 250-23(a) and 250-61(b)] This, in essence, prevents using the grounded system or service conductor for grounding equipment to it downstream from the service disconnecting means. Six exceptions to the general rule are provided. They are as follows:

Exception No. 1: A grounding electrode conductor shall be connected to the grounded conductor of a separately derived system in accordance with the provisions of Section 250-26(b). See Chapter Twelve of this text.

Exception No. 2: A grounding conductor connection shall be made at each separate building where required by Section 250-24. See Chapter Thirteen of this text.

Exception No. 3: For ranges, counter-mounted cooking units, wall-mounted ovens, clothes dryers, and meter enclosures as permitted by Section 250-61. See Chapter Ten of this text.

Exception No. 4: For services that are dual fed (double ended) in a common enclosure or grouped together in separate enclosures and employing a secondary tie, a single grounding electrode connection to the tie point of the grounded circuit conductors from each power source shall be permitted.

Exception No. 5: Where the main bonding jumper specified in Sections 250-53(b) and 250-79 is a wire or busbar, and is installed from the neutral bar or bus to the equipment grounding terminal bar or bus in the service equipment, the grounding electrode conductor shall be permitted to be connected to the equipment grounding terminal bar or bus to which the main bonding jumper is connected. See Chapter Fifteen of this text for information on equipment ground fault protection systems.

Grounding Equipment
by Grounded Circuit Conductor

Separately derived systems.
Section 250-23(a) Ex. No. 1

Each separate building where
required by Section 250-24.
Section 250-23(a) Ex. No. 2

Ranges, counter-mounted,
cooking units, wall-mounted
ovens, clothes dryers and
meter enclosures as
permitted by Section 250-61.
Section 250-23(a) Ex. No. 3

Dual fed (double-ended)
services with a secondary tie,
a single grounding electrode
connection is permitted.
Section 250-23(a) Ex. No. 4

Grounding electrode
conductor permitted to be
connected to equipment
grounding terminal bar under
specified conditions.
Section 250-23(a) Ex. No. 5

As covered in Section 250-27
for high-impedance grounded
neutral systems.
Section 250-23(a) Ex. No. 6

Load-side meter enclosures
that are near the service
disconnecting means where
service ground-fault
protection is not provided.
Section 250-61 Ex. No. 2

As required for electrode-type
boilers in Article 710, Part G.
Section 250-61 Ex. No. 4

Figure 9-10

Exception No. 6: As covered in Section 250-27 for high-impedance grounded neutral systems grounding connection requirements. See Chapter Four of this text.

The Code requires that where a secondary alternating-current system is to be grounded, the connection shall be made on the supply side of the service disconnecting means. It is further required that each secondary distribution system which is grounded shall have at least one additional connection to a grounding electrode at the transformer or elsewhere. That requirement applies to a secondary distribution system that is grounded and not to a separately derived system that is grounded. (Section 250-26)

The Code further permits the use of the grounded conductor on the supply side of the service disconnecting means for grounding meter housing and service equipment.

The requirement in Section 250-61, which prohibits grounding the grounded circuit conductor anywhere on the load side of the service disconnecting means, carries four exceptions which are as follows:

Exception No. 1. Except as Provided for in Section 250-24. This section refers to two or more buildings supplied by a single service and permits grounding of the grounded circuit conductor to a grounding electrode at the other buildings or structures.

Exception No. 2. Load-side meter enclosures may be connected to the grounded circuit conductor if no service ground-fault protection is installed and the meter enclosure is located near the service disconnecting means.

Exception No. 3. As provided in Section 250-60. Section 250-60 concerns the grounding of the frames of electric ranges, electric clothes dryers, wall-mounted ovens and counter-mounted cooking units. It is permitted, for existing installations only, to ground any of the frames of these appliances to the load side of the grounded conductor, provided the grounded circuit conductor is not smaller than No. 10 copper. New installations must be grounded by installation of an equipment grounding conductor. See Section 260-60.

Exception No. 4. As required for electrode-type boilers in Article 710, Part G. The neutral is commonly bonded to the electrode boiler vessel, especially where over 600 volt systems are used.

In accordance with Section 250-61(b) of the Code, grounding to the grounded conductor is not permitted on the load side of the service disconnect except for the special conditions cited above. That means that the grounded conductor on the load side of the service must be an insulated conductor and cannot be bare.

Underground metal-sheathed cable system.

The Code recognizes that if an underground service originates from a continuous underground metal sheath cable system and the sheath or armor is metallically connected to the underground system, or if an underground service conduit is used which contains a metal-sheathed cable which is itself bonded to the underground system, then the service run need not be grounded at the building and may be insulated from the interior conduit or piping. (Section 230-55)

The Code here recognizes that since the service armor is already bonded to a continuous underground metal sheath cable system, we have a path for fault-currents in the service cable so there would be no necessity to bond back to the service equipment in order to find a return path to the neutral.

Chapter Nine: The questions included here were developed using material included in this chapter. The answers can be found by reviewing the text. It is also important that students make use of the 1996 NEC®, where many answers can be found. See pages 236 - 237 for answers.

1. The conductor used to connect the noncurrent-carrying metal parts of equipment, raceways, and other enclosures to the system grounded conductor and/or the grounding electrode conductor at the service equipment, or at the source of a separately derived system is defined as ____.
   a. equipment grounding conductor
   b. main bonding jumper
   c. grounding systems conductor
   d. circuit bonding jumper

2. An equipment grounding conductor is intended to prevent an objectionable voltage above ground on conductor and equipment enclosures and to provide a low-impedance path for fault-currents. This path must also ____.
   a. be permanent and continuous
   b. have ample capacity to conduct safely any currents likely to be imposed on it
   c. be of the lowest practical impedance
   d. all of the above

3. The equipment grounding conductor or path must extend from the ____ point on the circuit to the service equipment where it is connected to the grounded conductor.
   a. closest
   b. service
   c. furthermost
   d. bonding

4. Where the overcurrent protection ahead of a copper branch circuit or feeder is sized at 225 amperes, the minimum size of the equipment grounding conductor is to be a No. ____.
   a. 4 copper
   b. 6 copper
   c. 8 copper
   d. 10 copper

5. The recognized materials for a raceway used as an equipment grounding conductor include ALL BUT ONE of the following ____.
   a. a conductor of copper or other corrosion-resistant material such as aluminum
   b. rigid or intermediate metal conduit
   c. electrical metallic tubing
   d. approved surface metal raceways

6. Listed flexible metal conduit and listed flexible metallic tubing are permitted to be used for equipment grounding purposes. Which one of the following is NOT true? ____
   a. Total of combined ground return in the same path cannot exceed 6 feet
   b. Must be terminated with fittings listed for grounding
   c: Not used on a circuit exceeding 15 amperes
   d. Not used on a circuit exceeding 20 amperes

7. Listed liquidtight flexible metal conduit in the ¾-inch through 1¼-inch trade size is to be used as an equipment grounding conductor. Which of the following statements is NOT true? ____
   a. The total length of the combined ground return in the same path cannot exceed 6 feet.
   b. Listed fittings must be used for grounding at connections.
   c. The circuit is permitted to be protected by a 100 ampere or less overcurrent device.
   d. The circuit is permitted to be protected by a 60 ampere or less overcurrent device.

8. Where rigid metal conduit is used as an equipment grounding conductor, all joints and fittings are required to be ____.
   a. approved
   b. tested
   c. made up wrenchtight
   d. sealed

9. For the grounding of portable or pendant equipment and where protected by fuses or circuit breakers rated or set at not over ____ amperes, the Code permits the use of a No. 18 copper wire as an equipment grounding conductor provided it is a part of a listed flexible cord assembly.
   a. 25
   b. 20
   c. 30
   d. 35

10. Under what conditions can building steel serve as an equipment grounding conductor to ground equipment? ____
   a. Where it is effectively grounded.
   b. Never.
   c. Where approved.
   d. By special permission.

Enclosure and Equipment Grounding

Objectives

After studying this chapter, the reader will be able to understand

- General requirements and definitions for enclosure and equipment grounding.
- Grounding of fixed and specific equipment.
- Grounding of nonelectrical equipment.
- Methods of effective grounding and grounding of equipment.
- Grounding of metal enclosures and panelboards.
- Installation of grounding-type receptacles.
- Grounding of cord- and plug-connected equipment.
- Equipment spacing from lightning rods.

Definitions.

Bonding Jumper: "A reliable conductor to ensure the required electrical conductivity between metal parts required to be electrically connected."*

Bonding Jumper, Equipment: "The connection between two or more portions of the equipment grounding conductor."*

General requirements.

Both service conductor enclosures and other conductor enclosures, where of metal, are required to be grounded. (Sections 250-32 and 250-33) That does not mean connecting equipment to a grounding electrode but rather complying with the requirements of Section 250-51 where "Effective Grounding" is carefully outlined. It is important to realize that wherever the Code states, "shall be grounded" it means effectively grounded as spelled out in Section 250-51. There is no distinction in "effective grounding" whether we are dealing with a grounded or an ungrounded system. Where grounding is not effectively accomplished, the situation, while bad in an ungrounded system, becomes worse in a grounded system.

Equipment grounding conductor.

It is important to recall that an equipment grounding conductor must be used for grounding equipment. This equipment grounding conductor is permitted to consist of any of the conductors or wiring methods identified in Section 250-91(b).

Installing a grounding electrode to ground equipment without having an equipment grounding conductor (in some cases the grounded service conductor) is unsafe and not permitted by the Code. See Sections 250-51 and 250-91(c).

Grounded Improperly

High Impedance Path

Figure 10-1

By referring to Figure 10-1, it can be seen that the grounding connections as shown comply literally with the wording of the Code in that the enclosures are grounded. But the wiring does not comply with Section 250-51 as it is not effectively grounded. In addition, the grounding shown in this figure violates Sections 250-51 and 250-91(c) as an earth return grounding circuit is indicated.

The wiring in Figure 10-2 complies literally with Sections 250-32 and 250-33 and also complies with Section 250-51 as an equipment grounding conductor is installed.

Grounded Properly

Low impedance path

High impedance path

Figure 10-2

It is obvious that we have a high-impedance fault current return path in Figure 10-1, while in Figure 10-2 we have a path having sufficiently low impedance to limit the voltage to ground and to facilitate the operation of the circuit protective devices in the circuit. Note that in Figure 10-2 there are two paths for current to return to the source. The primary and low impedance path is over the grounded neutral conductor while a second, high-impedance path in parallel with the grounded service conductor is through the grounding electrodes and the earth.

Service raceways and enclosures.

We have dealt in an earlier chapter of this text with the requirement for bonding service raceways and equipment. By bonding the service equipment to the grounded service and grounding electrode, we have complied with the requirements of Section 250-32.

*Reprinted with permission from NFPA 70-1996, National Electrical Code®, Copyright© 1995, National Fire Protection Association, Quincy, MA 02269. This reprinted material is not the official position of the National Fire Protection Association which is represented only by the standard in its entirety.

Metal Elbows Used Underground

Figure 10-3

An exception has been added to this section which exempts a metal elbow from the requirement that it be grounded where it is installed in an underground run of rigid nonmetallic conduit and is isolated from possible contact by a minimum cover of 18 inches to any part of the elbow. These metal elbows are often referred to as "pulling elbows" and are commonly installed in duct banks or other underground runs of rigid nonmetallic conduit and are more durable than PVC elbows during the cable pulling process.

A similar exception for other than service raceways has been added to Section 250-33.

Exceptions from grounding.

For other than service conductor enclosures, four exceptions are provided from the requirement that metallic conductor enclosures be grounded. (Section 250-33).

Exception No. 1 covers the metal enclosures for conductors added to existing installations of open wire, knob-and-tube wiring, and nonmetallic-sheathed cable which do not provide an equipment ground. Conditions that must be met are as follows: the raceway be less than 25 feet long, be free from probable contact with ground or a grounded object and is guarded against contact by persons.

Exception No. 2 exempts short sections of metal enclosures from grounding where used to protect cable assemblies from physical damage. No explanation is given in the Code for the meaning of "short sections" of metal enclosures. Since the standard length is 10 feet, "short sections" probably are less than 10 feet long.

Exception No. 3 excludes enclosures for certain remote control, signalling and fire protective sig-

nalling circuits that are not required to be grounded by Section 250-5(a).

Exception No. 4 permits a metal elbow to not be grounded where it is installed in an underground run of rigid nonmetallic conduit and is isolated from possible contact by a minimum cover of 18 inches to any part of the elbow. This provision is identical to that for services as provided in Section 250-32, Exception. See Figure 10-3.

Grounding of fixed equipment.

It is mandatory that noncurrent-carrying metal parts of fixed equipment, which are likely to become energized, be grounded under the six conditions cited in Section 250-42. Here, again, the term "shall be grounded" must not be interpreted literally to mean "connect to a grounding electrode" but must be interpreted in the light of all of Article 250 where effective grounding is outlined in Section 250-51. The Code does not define what is meant by "likely to become energized." Generally, if the equipment has exposed noncurrent-carrying metal parts and is supplied by electric current it should be grounded where any of the six conditions exist at the equipment.

The six conditions under which exposed non-current-carrying metal parts of equipment must be grounded are:

a. where within 8 feet vertically or 5 feet horizontally of ground or grounded metal objects and subject to contact by persons.

b. where located in a wet or damp location and not isolated.

c. where in electrical contact with metal.

d. where in hazardous (classified) locations as covered by Articles 500 through 517.

e. where supplied by a metal-clad, metal-sheathed, metal raceway, or other wiring method which provides an equipment ground. Exempted are short sections of metal enclosures covered in Section 250-33.

f. where equipment operates with any terminal at more than 150 volts to ground.

Four exceptions from the requirement that these equipments must be grounded are as follows:

Exception No. 1. Enclosures for switches or circuit breakers used for other than service equipment where accessible to qualified persons only;

Exception No. 2. Metal frames of electrically heated devices exempted by special permission (the written approval of the Authority Having Jurisdiction), in which case the frames shall be permanently and effectively insulated from ground;

Exception No. 3. Distribution apparatus such as transformers and capacitor cases that are mounted on wooden poles at a height of more than 8 feet from the ground or grade level; and

Exception No. 4. Double insulated equipment that is distinctively marked.

Grounding specific equipment.

The Code requires that exposed, noncurrent-carrying metal parts of certain specific equipment, regardless of voltage, shall be grounded. Those items are spelled out in Section 250-43 and include:
a. switchboard frames and structures supporting switching equipment.
b. generator and motor frames in an electrically-operated pipe organ unless the generator is effectively insulated from ground and from the motor driving it.
c. frames of stationary motors where:
   1. supplied by metal-enclosed wiring;
   2. in a wet location and not isolated or guarded;
   3. in a hazardous (classified) location; and
   4. it operates with any terminal at over 150 volts to ground. (Section 430-142)
d. Enclosures for motor controllers. Exempted are enclosures attached to ungrounded portable equipment and covers of snap switches that are lined.
e. Electrical equipment for elevators and cranes.
f. Electrical equipment in commercial garages, theaters and motion picture studios. Exempted are pendant lampholders supplied by circuits not over 150 volts to ground.
g. Electric signs, outline lighting and associated equipment as provided in Article 600. See Chapter Sixteen of this text for additional information on the subject.
h. Motion picture projection equipment.
i. Equipment supplied by Class 1 power-limited circuits and by Class 1, Class 2 and Class 3 remote-control and signalling circuits and by fire alarm circuits where these systems are required to be grounded by Part B of Article 250.
j. Lighting fixtures as required by Part E of Article 410 as follows:
   1. exposed conductive parts of lighting fixtures wired or attached to outlets supplied by a wiring method which provides an equipment ground; [Section 410-18(a)]
   2. fixtures that are supplied by a wiring method that does not provide an equipment ground must be made of insulating material and have no exposed conductive parts; [Section 410-18(b)] and
   3. fixtures with exposed metal parts must be provided wit h a means for connecting an equipment grounding conductor. (Section 410-20)
k. motor-operated water pumps including the submersible type. Also see Section 547-8(d) for specific requirements for Agricultural Buildings.
l. metal well casings must be bonded to the pump circuit equipment grounding conductor where a submersible type pump is installed.

Nonelectric equipment.

The grounding of nonelectric equipment is covered by the Code in Section 250-44. The equipment mentioned is considered as being likely to become energized and is thus required to be grounded as a safety measure. Included are:
(a) frames and tracks of electrically operated cranes and hoists;
(b) frames of nonelectrically driven elevator cars to which electric conductors are attached.
(c) hand-operated metal shifting ropes or cables of electric elevators,
(d) metal partitions, grill work and similar metal enclosures around equipment 1,000 volts and more between conductors. Exempted are substations or vaults under exclusive control of the electric utility; and
(e) mobile homes and recreational vehicles as provided in Articles 550 and 551.

The Fine Print Note following this section recommends that where extensive metal in or on buildings may become energized and is subject to personal contact, adequate bonding and grounding will provide additional safety.

Methods of grounding.

To provide the effective grounding required by the Code, it is important to connect the equipment grounding conductor recognized in Section 250-91(b) to the equipment in such a manner that the requirements for "effective grounding" are met.

For equipment that is fastened in place or connected by permanent wiring methods, the equip-

---

Connection of Equipment Grounding Conductor

- **Raceway - by fittings designed for the purpose - made up wrenchtight.**

- **Conductors - by exothermic welding or by listed pressure connectors or clamps.**

---

ment grounding conductor must be connected to the enclosure in a proper manner.

Grounding conductors and bonding jumpers must be connected to equipment that is required to be grounded by exothermic welding, listed pressure connectors, listed clamps or other listed means. (Section 250-113)

The only connecting means not required to be listed is the exothermic welding method. Since the welding is performed in the field, it obviously can't be listed. It is most important that equipment to be welded be clean and dry and that manufacturer's instructions be followed to ensure a satisfactory connection. These welds must be examined and tested after completion to be certain that a reliable connection has been made. It is common to test the welds in the field by x-ray where required by the job specifications or by striking the weld with a hammer after it has cooled for less demanding installations.

In the case of a metallic raceway being used as the equipment grounding conductor, the raceway must be connected to the enclosure by using fittings designed for the purpose. All connections must be made up wrenchtight using proper tools. This includes locknuts, bushings, conduit and electrical metallic tubing couplings and connectors. [Section 250-92(c)(1)]

Solder connections.

Connections that depend solely upon solder cannot be used for grounding connections.

The reason for this prohibition is that when equipment grounding conductors carry fault-current, they can get very hot. This elevated temperature may exceed the melting point of the solder and weaken or destroy the connection. This may create a hazard by leaving equipment at a dangerous potential above ground, creating a shock hazard to those who may contact the equipment.

It is permissible to secure the connections me-

chanically and then apply solder to the joint to make the electrical connection.

Special provisions for grounding certain appliances.

Special requirements are set forth in Section 250-60 for the grounding of frames of electric ranges, electric clothes dryers and similar appliances. These provisions apply to only existing installations. New installations must comply with the requirements for an insulated neutral as well as an equipment grounding conductor given in Sections 250-57 and 250-59.

Appliances or equipment to which Section 250-60 applies include:

(a) electric ranges;
(b) wall-mounted ovens;
(c) counter-mounted cooking units;
(d) clothes dryers, and
(e) outlet or junction boxes that are part of the circuit for these appliances.

**Wiring Methods for Appliances
Permitted to be Grounded by the Grounded Circuit
Conductor**

Figure 10-4

These appliances and equipment, such as the junction boxes in the circuit, are permitted to be grounded in two ways, either by use of an equipment grounding conductor or, except for mobile homes and recreational vehicles, by the use of the grounded circuit (neutral) conductor, provided all the conditions of this section are complied with. If a 3-wire with ground circuit is run, simply use the equipment grounding conductor for grounding metal equipment and install a grounding type receptacle.

The conditions for grounding the equipment to the grounded circuit conductor in existing installations that must be met are as follows:

(a) The appliances are supplied by a 120/240-volt, single-phase, 3-wire circuit or by a 3-wire circuit derived from a 208Y/120-volt, 3-phase, 4-wire, wye-connected system.

(b) The grounded (neutral) conductor is No. 10 copper or No. 8 aluminum or larger.

(c) The grounded (neutral) conductor is insulated. The grounded (neutral) conductor is permitted to be uninsulated if it is part of a type SE service-entrance cable and it originates at the service equipment, and

(d) Grounding contacts of receptacles that are a part of the appliance are bonded to the equipment.

It is important to note that all of these special conditions must be met before the appliances, and outlet and junction boxes that are a part of the circuit to the appliances are permitted to be grounded by use of the circuit neutral. Note that the supply cable must have an insulated neutral conductor where supplied from a sub-panel.

Use caution when applying the provisions of Section 250-60(c) which permits the use of a Type SE cable having a bare neutral for wiring these appliances when the circuit originates at service equipment. Section 338-3(b) no longer permits Type SE cable without an insulated neutral conductor to be used for the purposes of Section 250-60.

Wiring for new ranges and dryers.

New installations for ranges, dryers and similar appliances require a three-wire with ground cable having an insulated neutral conductor and an equipment grounding conductor. Receptacles, where installed, must be of the 3-pole, 4-wire grounding type. Supply cords, where used, must be of the 3-wire with ground type.

Neither the frame of the appliances nor the outlets or junction boxes that are a part of the supply to the appliances are permitted to be grounded to the circuit neutral. The frame of the appliances and junction boxes must be grounded by means of an equipment grounding conductor run with the branch circuit.

Care must be taken when these appliances are moved from a location employing one grounding scheme to a location having a different one. In the case where a 3-wire cable is used, the bonding jumper in the appliance junction box must be connected between the frame of the appliance and the neutral. Where a four-wire supply is used, the bonding jumper must be removed or disconnected, the neutral conductor be isolated from the appliance frame and the equipment grounding conductor connected to the frame.

Grounding New Range and Dryer Circuits

Range, dryer, etc. outlet box

Service Equipment

four conductors, neutral insulated

Figure 10-5

Outlet, device, pull and junction boxes.

Metal outlet, device, pull and junction boxes are required to be grounded in accordance with Article 250. See Section 300-9. It is important that the grounding of these enclosures be accomplished by the wiring method that supplies the enclosure. (Section 250-57) Under no circumstances should a metal enclosure be connected to a local grounding electrode in lieu of grounding them by means of the suitable wiring method or equipment grounding conductor unless specifically permitted by Code rules.

The metal raceway containing circuit conductors is permitted to be used for grounding these enclosures if installed properly with all connections made up wrenchtight. In addition, where fittings are used for connecting cables or raceways being used for grounding, the fittings used for connecting the wiring methods must be suitable for grounding.

Equipment grounding conductors that are supplied for grounding these enclosures must be sized in accordance with Table 250-95 based upon the rating of the overcurrent device protecting the circuit conductors in the raceway.

Section 370-40(d) requires that a means be provided by the manufacturer in each metal box that is designed for use with nonmetallic raceways and nonmetallic cable systems for the connection of the equipment grounding conductor.

Grounding of panelboards.

Section 215-6 requires that "where a feeder supplies branch circuits in which equipment grounding conductors are required, the feeder shall include or provide a grounding means in accordance with the provisions of Section 250-57 to which the equipment grounding conductors of the branch circuit shall be connected."

Isolated Neutral in Subpanel

Figure 10-6

Specific requirements for grounding panelboards are contained in Section 384-20. All panelboard cabinets and panelboard frames, if of metal, are required to be in physical contact with each other and must be grounded in accordance with Article 250 or Section 384-3(c). This section requires that where used as service equipment, the panelboard must be provided with a main bonding jumper located inside the cabinet for the purpose of bonding the enclosure to the grounded service conductor.

Where the panelboard is used with nonmetallic raceway or cable or with separate grounding con-

ductors, a terminal bar for terminating the equipment grounding conductor must be secured inside the cabinet. This terminal bar must be bonded to the cabinet if it is metal. If the enclosure is made of nonmetallic material, the terminal bar must be connected to the equipment grounding conductor that supplies the panelboard. Usually, the manufacturer provides matching and tapped holes along with appropriate screws or bolts for attaching the bar to the enclosure.

Panelboard with Nonmetallic Conduit Supply

Figure 10-7

Grounding and equipment grounding conductors are not permitted to be connected to the terminal bar for grounded (may be a neutral) conductors unless the terminal bar is identified for the purpose. This identification will usually be located on the manufacturer's label that is located within the cabinet. Another requirement is that the panelboard is used at a location where the grounded conductor terminal (neutral) bar is connected to a grounding electrode, as permitted or required by Article 250. These locations include at services, at the building disconnecting means (Section 225-8) and for separately derived systems (Section 250-26).

An exception permits an insulated equipment grounding conductor for an isolated grounding scheme to pass through the panelboard without being connected to the equipment grounding terminal bar in compliance with Section 250-74 Exception No. 4.

Generally, the grounded (may be a neutral) conductor is isolated from the enclosure downstream from the service equipment. Often, the term "floating neutral" is used to describe the grounded conductor's relationship to the enclosure as it is isolated electrically from the enclosure. Section 250-23(a) generally prohibits a grounding connection to a grounded conductor (may be a neutral) on the load side of the service disconnecting means. An exception is provided for separately derived systems and at separate buildings, as well as for certain appliances.

Isolated Grounding Receptacles

Figure 10-8

See Chapter 12 of this text for additional information on separately derived systems and Chapter 13 for grounding at more than one building on the premises.

Grounding-type receptacles.

Where grounding-type receptacles are installed, the grounding terminal of the receptacle must be connected to an equipment grounding conductor of the circuit supplying the receptacle. [Section 210-7(c)] Where more than one equipment grounding conductor enters a box, they must be connected together using a suitable and listed connector. Also, it is permitted to connect each of the equipment grounding conductors to the metal box individually using a listed clip or screw. A jumper is used to the receptacle so grounding continuity is not disturbed if the device is removed.

A grounding jumper from the receptacle to the box is not required where a device with listed grounding means is installed in a metal box that is properly grounded.

See the requirements in Chapter 8 of this text for additional information on this subject.

Isolated grounding receptacles.

Receptacles that have the equipment grounding terminal isolated from the mounting strap, and thus from the box, are commonly installed at computer terminals and cash registers. This is permitted by Section 250-74 Exception No. 4 for the purpose of reducing electrical noise (electromagnetic interference). The grounding terminal of the receptacle must be grounded by means of an insulated equipment grounding conductor that is run with the circuit conductors.

This insulated equipment grounding conductor is permitted to pass through one or more panelboards without connection to the terminal bar within the panelboard on its way back to, usually, the service disconnecting means. See Sections 250-114 Exception and 384-20 Exception. However, the insulated equipment grounding conductor must terminate within the same building at the building disconnecting means or source of a separately derived system. See Chapter 12 of this text for additional information on the subject of grounding separately derived systems.

Isolated equipment grounding.

The exception to Section 250-75 permits an equipment enclosure supplied by a branch circuit to be grounded by means of an insulated equipment grounding conductor contained within the raceway with the branch circuit that supplies the equipment.

Isolated Equipment Grounding

Comply with Section 250-74 Ex. No. 4
Supplied by only branch circuits.

Insulated equipment grounding conductor through raceway.

Listed nonmetallic raceway fitting required.
Required for reduction of electrical noise.

Figure 10-9

A listed isolating raceway fitting at the point of connection to the equipment must be installed.

Underwriters Laboratories performed tests of a similar grounding scheme to determine whether isolating the metal conduit from the equipment had an adverse effect on the grounding circuit impedance. They found no appreciable increase in impedance with the metal conduit isolated from the equipment and being grounded by means of the insulated equipment grounding conductor.

Short sections of raceway.

Where isolated sections of metal raceway or cable armor are required to be grounded, the Code requires in Section 250-56 that grounding of such sections be performed in accordance with the requirements of fixed equipment found in Section 250-57. The provisions here are that such sections of metal raceway or fixed equipment, where metallically connected to grounded cable armor or a metal raceway, are considered to be grounded by that connection since the grounded cable armor or metal raceway is a part of the equipment grounding conductor.

If the equipment is not connected to a metal raceway, then it is required that a grounding conductor be run with the circuit conductors. Such a conductor may be uninsulated, but if it does have an individual covering, the requirement is that the covering shall be of green color or green with one or more yellow stripes.

Another method of grounding where the equipment is not metallically connected to a metal raceway is by the use of a separate grounding conductor installed the same as a grounding conductor for conduit and the like. While the words are different here than were expressed in the first method, the end result is the same. This is because a separate equipment grounding conductor must also be run with the circuit conductors if we are to maintain the lowest practical impedance.

Grounding of cord- and plug-connected equipment.

Section 250-45 covers grounding of equipment connected by cord and plug. It is mandatory under certain conditions that noncurrent-carrying metal parts of cord- and plug-connected equipment, which are liable to become energized, be grounded. Specifically, equipment must be grounded where located in

   (a) hazardous (classified) locations,
   (b) if the equipment operates at more than 150 volts to ground. Exempted are;
      (l) motors, where guarded,

   (2) metal frames of electrically-heated appliances exempted by special permission in which case the frames shall be permanently and effectively insulated from ground, and
   (3) double insulated office equipment that is distinctively marked.

The Code cites in Section 250-45(c) specific equipment of the cord- and plug-connected type which must be grounded in residential occupancies. In addition, Section 250-45(d) lists specific cord- and plug-connected equipment in other than residential occupancies that must be grounded. In this section, recognition is given to the use of a listed system of double insulation which would provide the safety required without the use of grounding.

Note that the equipment grounding conductor must be sized in accordance with Table 250-95.

In other than residential occupancies, an exception is provided for cord- and plug-connected equipment that is supplied through an isolating transformer with an ungrounded secondary of not over 50 volts.

Spacing from lightning rods.

Section 250-46 requires that metal raceways, enclosures, frames and other noncurrent-carrying metal parts of electric equipment shall, wherever practicable, be kept at least 6 feet away from lightning rod conductors. If it is not practicable to secure the 6-foot separation, the Code requires the bonding together of the equipment and the lightning rod conductor at the point where the separation distance is less than 6 feet.

Spacing From Lightning System Conductors

Metal raceways, enclosures, frames, and other non-current-carrying metal parts of electric equipment shall be kept at least 6 feet away from lightning rod conductors, or they shall be bonded to the lightning rod conductors at the location where the separation is less than 6 feet.

Figure 10-10

Chapter Ten: The questions included here were developed using material included in this chapter. The answers can be found by reviewing the text. It is also important that students make use of the 1996 NEC®, where many answers can be found. See page 237 for answers.

1. A reliable conductor to assure the required electrical conductivity between metal parts required to be electrically connected is defined as a ____.
   a. grounding electrode conductor
   b. grounded conductor
   c. bonding jumper
   d. identified conductor

2. The connection between two or more portions of the equipment grounding conductor is defined as ____.
   a. equipment bonding jumper
   b. grounded
   c. neutral
   d. main bonding jumper

3. Exposed noncurrent-carrying metal parts of equipment must be grounded where within____ feet vertically or ____ feet horizontally of ground or grounded metal objects, and subject to contact by persons.
   a. 9 - 6
   b. 8 - 5
   c. 8 - 8
   d. 9 - 9

4. Exposed noncurrent-carrying metal parts of equipment are required to be grounded where located in wet or damp locations and are not ____.
   a. guarded
   b. identified
   c. shielded
   d. isolated

5. Where other than short sections of metal enclosures are installed as per Section 250-33, exposed noncurrent-carrying metal parts of equipment are required to be grounded where supplied by any of the following wiring methods EXCEPT ____.
   a. metal-clad, metal-sheathed cables
   b. knob-and-tube wiring
   c. other wiring methods that provide an equipment ground
   d. approved metal wireways

6. Exposed noncurrent-carrying metal parts of equipment not required to be grounded include enclosures for switches or circuit breakers used for other than ____ equipment where accessible to qualified persons only.
   a. welding
   b. service
   c. meters
   d. emergency

7. Exposed noncurrent-carrying metal parts of equipment not required to be grounded include metal frames of electrically heated devices unless exempted by ____ in which case the frames must be permanently and effectively insulated from ground.
   a. the product instructions
   b. the local government
   c. special permission
   d. the code

8. Exposed noncurrent-carrying metal parts of equipment not required to be grounded include distribution apparatus, such as transformers and capacitor cases, that are mounted on wooden poles at a height of more than ____ feet from the ground or grade level.
   a. 6
   b. 8
   c. 5
   d. 7

9. Exposed noncurrent-carrying metal parts of equipment not required to be grounded include ____ equipment that is distinctively marked.
   a. hospital
   b. triple insulated
   c. single insulated
   d. double insulated

10. Where added to existing installations of open wire, knob-and-tube wiring, and nonmetallic-sheathed cable without an equipment grounding conductor, a metal enclosure for conductors run in lengths not to exceed 25 feet, free from probable contact with ground, grounded metal, metal lath, or other conductive material, and if guarded against contact by persons are ____.
    a. required to be grounded
    b. not required to be grounded
    c. required to be protected by a GFCI
    d. not permitted unless approved

11. Motor-operated water pumps including the submersible type are required to be grounded when they operate at ____.
    a. any voltage
    b. 120 volts
    c. 240 volts
    d. 208 volts

12. All of the connecting means included below are required to be listed by a qualified electrical testing lab EXCEPT ____.
   a. pressure connectors
   b. lugs
   c. clamps
   d. exothermic welding

13. Connections that depend solely upon solder for grounding connections ____.
   a. are only permitted when approved
   b. cannot be used
   c. are permitted on the load side of the service
   d. are permitted on the line side of the service

14. Metal raceways that enclose service conductors are required to be grounded unless it is:
   a. installed on a pole
   b. more than 8 feet above the ground
   c. an elbow in PVC run covered by 18 in. of earth
   d. buried at a depth given in Table 300-5

15. All the following statements about grounding the frame of ranges and dryers are true EXCEPT:
   a. A three-wire with ground circuit is required for existing installations.
   b. The frame of the appliances is permitted to be grounded to the neutral in new installations.
   c. A two-wire with ground circuit is required for new installations.
   d. New installations must be grounded to comply with Sections 250-57 and 250-59.

16. All the following statements about grounding of panelboards supplied by a feeder in the same building as the service are true EXCEPT:
   a. The feeder must supply or provide the grounding means.
   b. The neutral conductor is permitted to ground the enclosure.
   c. The neutral conductor must connect to a neutral bar that is isolated from the enclosure.
   d. An equipment grounding conductor must be connected to an equipment grounding terminal bar.

Notes:

Chapter 11

Clearing Ground Faults and Short Circuits

Objectives

After studying this chapter, the student will be able to understand

- Ground faults and circuit impedance.
- Fundamentals of equipment grounding, circuit design, and test procedures.
- Common elements in clearing ground faults and short circuits.
- Sizing of equipment grounding conductors.
- Purposes served by neutral on grounded systems.
- Conductor withstand ratings.

Ground faults and circuit impedance.

Section 310-2(a) requires, generally, that conductors be insulated. The exception allows covered or bare conductors to be used where specifically permitted in the Code. Ungrounded or hot conductors must be insulated for the applied voltage. Typical voltage rating of conductors is 300, 600, 2000, 5000 and 15,000. Conductors are available with much higher rated insulations, although these are primarily used for distribution of electrical energy.

Table 310-13 gives conductor application and insulations. It gives the trade name, type letter such as MI, maximum operating temperature and application provisions. It also gives the conductor insulation, size in AWG (American Wire Gage), thickness in mils and outer covering, if any.

Bare or covered conductors are permitted to be used in certain cases. Grounded service-lateral conductors are permitted to be used bare under the conditions given in Section 230-30. Bare service-entrance conductors are permitted to be used as provided in Section 230-41. Of course, insulated, covered or bare equipment grounding conductors are permitted to be used by Section 250-91(b). Some Code sections require insulated conductors for specific applications.

If no hazard can exist on a distribution system unless there is an insulation failure of the ungrounded or hot conductor, it follows that we should take every precaution to furnish the best possible insulation consistent with an economical installation. Good installation practices should be followed carefully to ensure that cable insulation is not damaged during the installation process. Cable insulation can be easily damaged by pulling operations if raceways are not cleaned prior to the installation of conductors, if excessive bends are in the run or where long or heavy pulls require the use of heavy-duty pulling equipment.

Further, if and when insulation does break down, the hazard can exist only as long as the circuit remains energized. Every effort should be made to deenergize, as quickly as is practical, any circuit on which a fault to ground has developed.

When first considered, the use of a copper or aluminum equipment grounding conductor may seem to offer an ideal method of getting a low-impedance path. It will in most cases, but that is not always true. Each individual circuit must be studied in relation to the size and length of the circuit, the size of its overcurrent device and the size of the conductor enclosure before any conclusion can be arrived at.

Fundamentals of equipment grounding circuit design.

Briefly, it can be stated that to get low impedance of the grounding system in an ac system, we must keep the circuit conductors and the equipment grounding conductor intimately together at all times. [Section 250-57(b)] The equipment grounding conductor may, of course, be a copper or aluminum conductor or the metal enclosure of the conductors, where acceptable by Code for such use.

In an ac electrical distribution system, whether grounded or ungrounded, inductive reactance exerts a powerful influence by directing the return current to flow in a path closely paralleling the outgoing power conductor. In addition to resistance, an inductive reactance value is associated with every conductor in an ac system. The inductive reactance (expressed in ohms) increases as the spacing between the conductors increases. This indicates that the inductive reactance of a current path closely paralleling the phase conductors offers a lower impedance to the ground-fault current than any other current path regardless of a lower resistance of the other current paths. Inductive reactance will be the predominate factor in determining current division in parallel ground return paths in heavy or larger circuit constructions. This has been proven in actual tests where it is shown that there is a strong tendency for ground return currents to seek a path physically close to the power conductor over which the outgoing current flows.[1]

All Conductors to be Together

Sections 250-57(b) and 300-3(b)

Figure 11-1

Usually, the conduit or metallic raceway that encloses the conductors provides an excellent fault return path. The presence of magnetic material in the power conductor enclosure (conduit or raceway) introduces additional inductive effects tending to confine the return ground currents within the magnetic enclosure. Installation of an external equipment grounding conductor is quite ineffective. Connections to nearby structural building steel members, in an effort to serve as an equipment grounding means, are equally ineffective as well.

The intentional or accidental omission of using the metallic conductor enclosure as an equipment grounding conductor may lead to large induced voltages in nearby metallic structures, which may appear as a dangerous shock hazard or unwanted circulating currents. Only by the installation of an internal equipment grounding, in parallel with the raceway, can the current carried by the raceway be reduced. Joints in conduit and raceways must be connected wrench-tight, using proper tools, for the raceway to function effectively as an equipment grounding conductor.

Test procedure.

As illustrated in Figure 11-2, a special installation of 2½-inch rigid steel conduit and 4/0 copper cable was made for this investigation. It was installed in a building previously used for short-circuit testing because the building had heavy steel column construction, and all columns were tied to an extensive grounding mat composed of 250 kcmil copper cables. The conduit was supported on insulators throughout the 100-foot length. The conduit was about 5 feet from a line of building columns. The external 4/0 cable was spaced about 1 foot from the conduit on the side opposite the building columns.

The setup was intended to simulate an electrical feeder circuit as may be found in many commercial and industrial plants. It allowed a study of a wide variety of equipment grounding arrangements. In every case the supply current was through the "A" cable run through the conduit. The "B" conductor served as an internal equipment grounding conductor. The "C" conductor was connected to the run of steel metal conduit, the "G" conductor was connected to the steel building columns and the "H"

Grounding Current Test Procedure

Steel building columns

Copper strap

G

100 feet

A

B   2-1/2-inch Steel Conduit

1 ft.

C

H   All cables - 4/0 Copper

Figure 11-2

conductor was the external equipment grounding conductor spaced about 1 foot away from and parallel to the conduit. With this arrangement it was possible to simulate various fault conditions at the right end and to verify various fault current return paths.

One series of tests was made at low current of 200 and 350 amperes using an ac welding transformer as a source of 60 cycle power. At these low-current magnitudes, the current could be maintained for extended periods. Voltage measurements were made with high quality indicating meters. Current measurements were made with a clip-on ammeter.

A second series of tests was made at high current, around 10,000 amperes using a 450 kVA, 3-phase, 60 cycle transformer with a 600 volt secondary as a source of power. Switching was done at the 13,800 primary voltage. An induction relay was used to control the duration of current flow to about ¼ second. An oscillograph was used for all measurements of current and voltage.

Tests performed, low-current.

A1 - 350 amperes out on internal conductor (A), return on internal conductor (B).

A2 - 350 amperes out on internal conductor (A), return on conduit (C).

A3 - 200 amperes out on internal conductor (A), return on conduit (C).

A4 - 350 amperes out on internal conductor (A),

return on conduit (C) and external conductor (H).

A5 - 200 amperes out on internal conductor (A), return on conduit (C) and external conductor (H).

A6 - 350 amperes out on internal conductor (A), return on conduit (C) and building steel (G).

A7 - 200 amperes out on internal conductor (A), return on conduit (C) and building steel (G).

A8 - 350 amperes out on internal conductor (A), return on conduit (C) and internal equipment grounding conductor (B).

A9 - 200 amperes out on internal conductor (A), return on conduit (C) and internal equipment grounding conductor (B).

A10 - 350 amperes out on internal conductor (A), return on building steel (G) and external equipment grounding conductor (H).

A11 - 200 amperes out on internal conductor (A), return on building steel (G) and external equipment grounding conductor (H).

Tests performed, high current.

B2 - 11,200 amperes out on internal conductor (A), return on conduit (C).

B3 - 11,070 amperes out on internal conductor (A), return on conduit (C).

B4 - 11,070 amperes out on internal conductor (A), return on conduit (C) and external equipment grounding conductor (H).

B5 - 11,080 amperes out on internal conductor (A), return on conduit (C) and external equipment grounding conductor (H).

B6 - 10,830 amperes out on internal conductor (A), return on conduit (C) and building steel (G).

B7 - 10,910 amperes out on internal conductor (A), return on conduit (C) and building steel (G).

B8 - 11,620 amperes out on internal conductor (A), return on conduit (C) and internal equipment grounding conductor (B).

B9 - 11,380 amperes out on internal conductor (A), return on conduit (C) and internal equipment grounding conductor (B).

B10 - 8,710 amperes out on internal conductor (A), return on building steel (G) and external equipment grounding conductor (H).

Results.

The results of the tests performed are shown in Table 11-1. The test number is for reference purposes. The next two columns show the cable connections used to determine current flow. Next, the current values are shown, first the total input current through cable A, and next the return current through the conduit and its percentage of the total. The other columns with an "I" heading, along with a subscript indicating the circuit, show the amount of current returning over other possible paths.

Current Magnitudes

Test No.	Out On	Return On	IA Total	Ic Amperes	Per Cent of Total	I_B	I_H	I_G	E_{AC}	E_{CG}	E_{AB}	E_{AH}	E_{AG}	E_{GB}
	Current Flow			Ic					Voltage Magnitudes					
Low-Current Tests														
A1	A	B	350	0		350	0	0	2.47		4.85			
A2	A	C	350	350	100	0	0	0	15.9	0.45	2.5			
A3	A	C	200	200	100	0	0	0	9.05	0.15	1.51			
A4	A	CH	350	340	97	0	12	0	16.0	0.05	2.55			
A5	A	CH	200	190	95	0	8	0	9.13	nil	1.55			
A6	A	CG	350	340	97	0	0	12	14.6		2.54	14.4		
A7	A	CG	200	180	90	0	0	8	9.5		1.50	9.4		
A8	A	CB	350	62	18	290	0	0	4.55	nil	4.55			
A9	A	CB	200	40	20	150	0	0	2.68	nil	2.68			
A10	A	GH	350	0	0	0	160	160	14.0	12.5	2.5	26.4	26.4	24.6
A11	A	GH	200	0	0	0	98	98	9.2	8.1	1.5	17.1	17.1	15.1
High-Current Tests														
B2	A	C	11,200	11,200	100	0	0	0	168	36*				
B3	A	C	11,070	11,070	100	0	0	0	173	38*				
B4	A	CH	11,070	11,200	101	0	1,140	0	173	18*				
B5	A	CH	11,080	11,090	100	0	1,220	0	173	17*				
B6	A	CG	10,830	10,770	99	0	0	1,080	168		71			
B7	A	CG	10,910	10,780	99	0	0	1,145	173	9*				
B8	A	CB	11,620	5,810	50	5,660	0	0		27*				155
B9	A	CB	11,380	6,070	53	5,620	0	0	146	25*				
B10	A	CH	8,710	0	0	0	4,300	4,500	146					268

*Distorted wave shape. Tabulated values are crest / $\sqrt{2}$

Table 11-1. Measured Electrical Quantities.[2]

"No further analysis is needed to show conclusively that only by the use of an internal grounding conductor can any sizable fraction of the return current be diverted from the raceway. In spite of the extremely low resistance of the building structural frame, it was ineffective in reducing the magnitude of the return current in the conduit. See tests A6, A7, B6 and B7.

"Some interesting secondary effects were observed in the course of the tests. The first high-current test produced a shower of sparks from about half of the couplings in the conduit run. From one came a blowtorch stream of sparks which burned out many of the threads. Several small fires set in nearby combustible material would have been serious if not promptly extinguished.

"The conduit run had been installed by a crew regularly engaged in such work and they gave assurance that the joints had been pulled up to normal tightness and perhaps a little more. A short 4/0 copper (wire) jumper was bridged around this joint but, even so, some sparks continued to be expelled from this coupling on subsequent tests. Other couplings threw no more sparks during subsequent tests. Apparently, small tack welds had occurred on the first test.

"In one high-current test, the conduit termination was altered to simulate a connection to a steel cabinet or junction box (see Figure 11-3). The bushing was applied finger-tight. In one test, with about 11,000 amperes flowing for about ¼ second, a fan-shaped shower of sparks occurred parallel to the plate. In the process, a weld resulted and the parts were separated only with considerable difficulty, with the use of wrenches and a hammer. This suggested that a repeat shot (of current) would have produced no disturbance (shower of sparks).

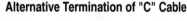

Alternative Termination of "C" Cable

Bushing

Locknut

Steel plate

"C" Cable

Figure 11-3

"During high-current test B10 (conduit circuit open), a shower of sparks was observed at an intermediate building column. Careful inspection disclosed that the origin was at a spot at which a water pipe passed through an opening cut in the web of the steel beam involved. Here is evidence of the objectionable effects of forcing the short-circuit current to seek return paths remote from the outgoing conductor. The large spacing between outgoing and returning current creates a powerful magnetic field which extends far out in space around the current-carrying conductors."[2]

Circuit analysis.

"The reactance of the circuit including the 'B' conductor will be the lowest. Next will be the innermost tube of the conduit, followed by others in successive order until the outer tube is reached. The inductance of these tubular elements of the steel conduit assumes unusual importance because of the high magnetic permeability. Next in spacing (impedance) is the external grounding conductor (H cable) and last, the structural members of the building frame and their interconnecting grounding cables buried below floor level."[2]

"In test B10, both the exterior grounding conductor (H cable) and the building frame (G terminal) were connected to provide parallel paths for the return current, but the conduit circuit was left open (C terminal not connected). The test results clearly evidence the powerful forces tending to maintain current flow in the conduit circuit. Note that across the open connections at the 'C' terminal, a voltage of 146 volts (or, more significantly, over 50 percent of the impressed driving voltage), is required to force the current to return via the 'H' cable and the building frame in parallel. Such a voltage could be a serious shock hazard.

"Furthermore, unless the conduit was well insulated throughout its entire length (which is usually impossible or impractical in typical commercial or industrial installations), there would be a significant number of sparks at stray points to constitute a serious fire hazard. It was during this test that a shower of sparks occurred between magnetic members in the building system that was caused simply by the strong magnetic field extending far out from the power conductors."[2]

Conclusions.

"The significance of this investigation points unmistakably to the conclusions presented earlier in this paper. Effective use of the conduit or raceway in the equipment-grounding system is paramount. Additional work is needed to develop joints which will not throw fire during faults. To improve effectiveness requires greater conductivity in the conductor enclosure or the use of an internal (equipment) grounding conductor. Grounding cables (grounding electrode conductors) connecting building structure with grounding electrodes (connection to earth) are needed to convey lightning currents or similar currents seeking a path to earth, but these conductors will play a negligible part in the performance of the equipment grounding system. Of course, the importance of proper equipment grounding becomes greater with the larger size feeder circuits and the availability of higher short-circuit currents."[2]

Common elements.

Clearing faults involves one or two parts, depending on whether the service is grounded or ungrounded. In a grounded distribution system we have two parts; (a) the grounded system conductor and (b) the equipment and conductor enclosure grounding system. In an ungrounded system, the grounding system covers only (b) the equipment and conductor enclosure grounding system.

Effective grounding.

To have effective grounding for both a grounded and an ungrounded system, the grounding system shall provide (1) a permanent and continuous path; (2) shall have ample carrying capacity to conduct safely any fault currents likely to be imposed on it; and (3) shall have impedance sufficiently low to limit the voltage to ground and to facilitate the operation of the circuit protective devices in the circuit. (Section 250-51.)

Impedance sufficiently low means, for all practical purposes, that the grounding path and the circuit conductors must always be within the same metallic enclosure, or the metallic enclosure under certain conditions may be used to provide the equipment grounding path for the circuit conductors within it.

Figure 11-4 courtesy of Square D

Overcurrent Device - Amps	Max Z of Ground-Fault Circuit in Ohms. Includes Fault Impedance.
60	0.4
100	0.24
200	0.12
400	0.06
600	0.04
800	0.03
1200	0.02
1600	0.015
2000	0.012
2500	0.0096
3000	0.008
4000	0.006

The Code gives no maximum impedance of the ground-fault circuit but states in Section 250-51 that it shall be sufficiently low to facilitate the operation of the circuit-protective devices. This path should have an impedance no greater than will allow the circuit breaker or fuse to reach its instantaneous pickup operating range. This will cause the overcurrent device to operate quickly to remove the fault from the system. Any fault current less than the instantaneous operating value will extend, by some value, the opening time of the overcurrent device.

Every manufacturer of circuit breakers and fuses publishes operating or trip curve charts for their products. See Figure 11-4 for one example. These charts should be carefully reviewed to be certain the ground-fault path has an impedance low enough to allow the overcurrent device to operate quickly to reduce thermal damage to the circuit and equipment. As can be seen by reviewing these charts, circuit breakers have a different trip curve depending on whether they are single-, double- or three-pole configurations. In addition, the same family of circuit breakers will have different trip curves for different ampere-rated circuit breakers. Look for the multiple of the circuit breaker rating at which it reaches its instantaneous pickup rating which is no time delay in the operation of the breaker.

A similar review of the fuse operating characteristics chart should be performed to select a ground-fault circuit that will result in instantaneous operation of the fuse.

Clearing short circuits.

The method employed for clearing a short circuit is the same regardless of whether the system is grounded or ungrounded. Essentially, it involves placing an overcurrent device in series with each ungrounded system conductor.

In the event of a "short circuit," which is a fault from conductor to conductor, the words "as quickly as is practical" means a very short period of time down to as low as a fraction of a cycle depending on the amount of short-circuit current that will flow and the characteristics of the overcurrent device. The high-interrupting capacity current-limiting fuse does that for us automatically within virtually all ranges of short-circuit current values. In addition current-limiting circuit breakers and circuit breaker current limiters are available from a wide variety of manufacturers. Where properly applied, these current-limiting devices significantly reduce the thermal and mechanical damage to electrical equipment.

Current-limiting fuses and their correct application are covered in Underwriters Laboratories Electrical Construction Materials Directory Guide Card (JCQR) information. There we find the following: "The term 'current limiting' indicates that a fuse, when tested on a circuit capable of delivering a specific short-circuit current (RMS amperes symmetrical) at rated voltage, will start to melt within 90 electrical degrees and will clear the circuit within 180 electrical degrees ($\frac{1}{2}$ cycle).

Because the time required for a fuse to melt is dependent on the available current of the circuit, a fuse which may be current limiting when subjected to a specific short-circuit current (rms amperes symmetrical) may not be current limiting on a circuit of lower maximum available current."

A current-limiting circuit breaker is defined in the Underwriters Laboratories Electrical Construction Materials Directory Guide Card (DIVQ) information as, "A current-limiting circuit breaker is one that does not employ a fusible element and when operating within its current-limiting range, limits the let-through I^2t (current squared time) to a value less than the I^2t of a $\frac{1}{2}$ cycle wave of the symmetrical prospective current. Current-limiting circuit breakers are marked 'current-limiting' and are marked either to indicate the let-through characteristics or to indicate where such information may be obtained."

Circuit breaker current limiters are covered in Underwriters Laboratories Electrical Construction Materials Directors in Guide Card (DIRW). "Circuit breaker current limiters are designed in

Clearing Short Circuits

Short circuits are cleared by the overcurrent device in series with the fault. Note the current flow in the circuit.

Figure 11-5

conjunction with specific circuit breakers and to be directly connected to the load terminals of the circuit breakers. They contain fusible elements which function only to increase the fault current interrupting ability of the combination which is intended for use in the same manner as circuit breakers when installed at the service and for branch circuit protection. The limiters are rated 600 volts or less."

It is vital to carefully follow manufacturer's instructions when applying current-limiting fuses or circuit breakers.

In some cases, depending on the available fault current, this equipment may not be current limiting at all. This is due to the "inverse time" nature of these overcurrent devices. "Inverse time" means that as the amount of current flowing through the device increases, the operating time of the device reduces. Overcurrent devices see fault current at lower operating ranges as a load and react much slower that at their "current-limiting" range.

Many installations of fuses or circuit breakers are installed in a series configuration, that is two or more fuses or circuit breakers in series for the circuit. These devices that have been tested for their operating compatibilities may be marked by the manufacturers with a "series-combination" rating. In this case, usually, the downstream overcurrent device is rated below the fault current that is available at its line terminals while the overcurrent device closest to the source is rated at or above the fault current that is available at its line terminals.

Where a series-rated system is installed, only the equipment that has been tested to determine its suitability can be installed in this manner. Suitability is determined by reference to

manufacturer's identification of components on decals located on the equipment. For additional information see "Interplay of Energies in Circuit Breaker and Fuse Combinations." [3]

By studying the time-current characteristic curve of such fuses and starting from a value of from five to six times the rating of the fuse, we will note a clearing time of about one second and a time of well below one cycle for values of 50 times fuse rating. A further study will indicate that such high-interrupting capacity current-limiting fuses have a pronounced current-limiting effect at values as high as fifty times the fuse rating.

Whether a fault current be a short circuit or a ground fault, the overcurrent devices ahead of the point of fault will ALL see the fault current. Placing an overcurrent protective device at the point the conductor receives its supply, as required by Section 240-21, will provide the short-circuit protection that is necessary. The rating of the overcurrent device is based upon Section 240-3. The basic requirement is that conductors be protected in accordance with their rated ampacity in accordance with Section 310-15.

Clearing ground faults.

Where it comes to clearing a "ground fault" two things control the flow of current, that is,

(1) The impedance of the circuit.
(2) The available short-circuit amperes of the system.

Where a ground fault occurs, however, we find that we are dealing with a different circuit than in the case of a short circuit. In a ground fault, we can, and do, get such a high impedance of the fault circuit that the controlling factor of current flow is the impedance of the ground-fault circuit. In this case, the only part the available short-circuit capacity plays is in its ability to maintain voltage. The result is that, for all practical purposes, the amount of current that will flow in a ground-fault circuit is not dependent on the available capacity of the system, other than its ability to maintain full voltage during a ground fault.

A relatively large voltage drop can be tolerated in this grounding conductor. The impedance of the complete ground-fault circuit, from the point of fault back to the transformer, should never be higher than would permit a minimum current flow of about 600 percent of the overcurrent device rating. (Review the operating characteristic chart for the overcurrent device for the correct value.)

Clearing Ground Faults

Figure 11-6

Clearing Faults in Service Equipment

The main bonding jumper is the vital link to clear a ground fault that may occur in service equipment.

Grounded neutral

Main bonding jumper

The grounding electrode has little effect in clearing a fault in the service.

Figure 11-7

This will provide a factor of safety to allow for the variable impedance at the point of fault. Even with a 2,000 ampere overcurrent protective device and a 12,000 ampere current flowing (600 percent), the clearing time will be approximately a little less than one second. This is a clearing time which may be considered satisfactory if the hazards are to be kept at a reasonable minimum.

Clearing faults in service equipment.

The service equipment enclosure is probably the most vulnerable for ground faults. There is no overcurrent protection on the line side of service conductors, only short-circuit protection provided by the utility's transformer primary overcurrent protection. Overload protection is provided by the service overcurrent device in series with the service-entrance conductors at the load end of the service-entrance conductors. This is the reason the Code requires the service disconnecting means to be located outside the building, or, if inside, nearest the point of entrance of the conductors into the building.

If, on the other hand, a ground fault does not develop on the load side of the service, but at a point on the line side of the service, then that fault

can only be cleared by the overcurrent device ahead of the transformers. This protection is the primary fuses for the transformer which can be many times the rating of the secondary overcurrent device.

In many cases, a ground fault on the line side of the service will not clear through the primary overcurrent devices and can only clear by developing into a short circuit or by burning itself clear. This is the reason Section 230-70(a) requires that the service disconnection means be located outside of a building or structure, or, if inside, nearest the point of entrance into the building. This limits the risk of having conductors with no short-circuit protection inside a building.

If the electrical equipment on the line side of the service is not properly bonded and a properly-sized main bonding jumper installed, then we are unlikely to pass enough current to clear the ground fault through the overcurrent protective devices on the line side of the utility transformer.

Analysis of clearing ground-faults.

Now let us take a simple system and analyze the conditions existing in the event of a ground fault. In the first case, the neutral is run from a grounded system where it is connected to the service equipment. In the second example, the grounded service conductor has not been run from the system to the service disconnecting means.

Purposes served by the neutral on a grounded system.

For maximum safety and to comply with Section 250-23(b), the neutral or grounded conductor must be run from a grounded system to all services and be bonded to each service disconnecting means enclosure even though all power may be utilized at line voltage only. This is required even though the service supplies only line-to-line loads. [Sections 250-23(b) and 250-53(b)]

The neutral of any grounded system serves two main purposes. First, it permits utilization of power at line-to-neutral voltage, and thus will serve as a current-carrying conductor to carry any unbalanced current back to the source. Second, it plays a vital part in providing a low-impedance path for the flow of fault currents to facilitate the operation of the overcurrent devices in the circuit, as required by Section 250-51.

The neutral provides the lowest impedance return path for fault currents to the transformer neutral point as can be traced in Figure 11-8. If the neutral is not needed for voltage requirements, it still must be run to the service and connected to the equipment grounding conductor at the service. In that event, the neutral conductor no longer serves as a neutral but as a grounded conductor. If that is not done, it is difficult, if not impossible, to clear a fault on the system.

Grounded (neutral) conductor installed.

In Figure 11-8, the neutral of the system serving as the grounded conductor is carried through to the service equipment and bonded to the equipment grounding conductor to provide a low-impedance path directly to the transformer. Here we see that a fault current is not required to go through the earth to complete the circuit, but will go through the grounded service conductor, which is a low impedance path. This will very likely permit sufficient current to flow to operate the overcurrent device. The parallel path through the grounding electrode conductor and the grounding electrode and the

earth still exists. The grounded service conductor, forming a relatively very low impedance path, will carry most of the fault current, probably 90 percent or better.

It is obvious, then, that to get all the protection afforded by a grounded system, the grounded conductor must be run to the service and must be bonded to the equipment grounding conductor even though the neutral is not required to serve any load whatever. For the same reason, where the neutral is needed for voltage requirements, the neutral size should be based not only on the neutral load demands, but on the basis of the service overcurrent device and the amount of fault current that must flow in order to operate the overcurrent devices.

Grounded Service Conductor Run to Service

Low impedance path

High resistance and impedance path

Figure 11-8

Neutral not installed.

Figure 11-9 shows a 120/240-volt single-phase grounded system where all the load is supplied at 240 volts. A grounding electrode conductor is properly connected to a low-resistance grounding electrode, the water supply system, and bonded to the equipment grounding conductor, all in accordance with the requirements of the Code (prior to the 1962 Code). However, the neutral is *not* carried into the service equipment since all power utilization is at 240 volts only. For serving the load, the neutral has no useful purpose, and it would at first appear as if it could be omitted.

From the standpoint of limiting the voltage between equipment and ground under normal conditions and from the standpoint of utilization of power at 240 volts, the circuit, as shown in Figure 11-9, will work perfectly, so long as the insulation remains perfect and no ground fault occurs on the system.

If, however, a ground fault should occur on the system in the equipment, as illustrated in Figure 11-9, owing to insulation failure, the voltage between equipment and ground will rise considerably.

Starting at the point of fault, we have first the impedance of the fault, then the conduit itself becomes a conductor in the fault-current circuit, then the grounding electrode conductor and the grounding electrode. However, the fault current must travel from the grounding electrode through the earth itself to the point where the neutral is grounded at the transformer, then through the transformer and back to the service and, finally, through the overcurrent device to the point of fault.

A study of the fault circuit as diagrammatically represented in Figure 11-9 shows that we would be unlikely to have an impedance of less than 22 ohms as the sum of all the impedances shown. That is an optimistically low figure at best. The maximum current that can flow in this circuit is 5.5 amperes for the 120/240 volt circuit shown. With a 100-ampere service and 20-ampere overcurrent device, it is obvious that the overcurrent device would not operate. A serious shock hazard, as well as fire hazard, will exist until the circuit is manually opened. If the circuit was opened as a result of being grounded properly, then the only period of time a shock hazard would exist would be for the duration of the fault. The fire hazard, as well as shock hazard, would be reduced to a relatively short period of time.

When a fault occurs on a system, it is only during the period while the fault exists that a potential hazard is present. It is, thus, of the utmost importance to safety that the fault clearing time be held to the shortest practical period of time.

With the wiring method as shown in Figure 11-9, the fault will not clear and may exist for minutes, hours or even days before it is recognized. Further, that recognition may come from observing a fire, or from noting that a victim has received a shock, which sometimes may even be fatal. It is a matter of record that many serious, and sometimes fatal, accidents have resulted from faults which were not cleared promptly because the system was not properly grounded.

If we were to assume the unlikely, but nevertheless possible, condition that the neutral at the transformer was grounded to the same water pipe system, then the resistance of the fault path would be appreciably decreased. Because of the wide separation between the service conductor and the water pipe, the reactance and, thus, the impedance of the fault circuit would remain high. The probabilities are that the fault current would not reach a high enough value to operate the overcurrent device. Again, fire and damage to equipment would continue until the circuit was manually opened. To improve the safety of such a system, a low-impedance path must be provided to pass enough current to clear the circuit through the overcurrent devices.

Three-phase services.

Identical reasoning to that for single-phase systems may be applied to any multi-wire, multiphase grounded system. Any 3-phase power supply taken from such a grounded transformer bank must have the neutral or grounded conductor brought into the service of the power supply to satisfy 250-23(b). This is true regardless of whether or not there is neutral load at the service. The grounded conductor provides a low-impedance path for fault current to return to the source.

Open system.

The above statements applying to maximum ground-fault currents would not apply to an open system. If there were no metallic enclosures, the impedance of the ground-fault circuit would be much lower. Accordingly, greater ground-fault currents may be expected in actual practice in an open or nonmetallic installation. However, since most systems of 600 volts or less are metal enclosed, we

Grounded Service Conductor Not Run to Service

High resistance and impedance path

Figure 11-9

may not get high ground-fault currents in such systems. For open systems, it is vital that the equipment grounding conductor be run with the circuit conductors to maintain a low impedance ground-fault path.

Recommended length of conduit for use as equipment grounding means.

The National Electrical Code® places no restriction on the size or length of rigid metal, intermediate metal conduit or electrical metallic tubing where used as an equipment grounding conductor. Independent tests have shown that consideration must be given to both size and length of conduit.

(Extensive work to determine the maximum safe length of conduit to serve as an equipment grounding conductor has recently been done by the School of Electrical and Computer Engineering at the Georgia Institute of Technology in Atlanta, Georgia.[4] Computer software has been developed which will allow the calculation of almost an unlimited combination of conductors and conduit for use in the ground-return path. This software can also be used for calculating the maximum length of equipment grounding conductors that are safe to use where not installed in metallic conduits.)

Where a metallic conductor enclosure is used as an equipment grounding conductor, it must have continuity as well as the required conductivity to pass enough current to facilitate the operation of the overcurrent devices. Some electrical inspection authorities require that an equipment grounding conductor be installed inside the conduit to account for poor workmanship of the raceway installation or to maintain continuity where fittings may be broken during use.

The engineer should examine the equipment grounding conductor (the metal enclosure) to be assured it will function properly in the event of a ground fault. Where wireways, auxiliary gutters and busways have steel enclosures, it will be found that there may be enough cross section to serve as an equipment grounding conductor. It is questioned if electrical connections between lengths is adequate for carrying enough fault current to clear the fault. These raceways may require a supplemental equipment grounding conductor.

The conductor enclosure, conduit, raceway, etc., also may be acceptable as the equipment grounding conductor in lieu of the copper conductors given in Table 250-95. However, it may be said, in general, that if a conduit, aluminum or steel, or electrical metallic tubing meets the Code requirements for conductor fill, the enclosure will provide an acceptable equipment grounding conductor.

In the case of the average busway up to 1500 amperes rating, there is enough steel in the enclosure to provide an acceptable equipment grounding conductor if proper conductivity is assured at the joints. For busways of higher rating, it is doubtful if the steel enclosure is heavy enough. For all sizes of busways, the matter of good electrical connection at the joints must be checked carefully. Many busways have aluminum enclosures. In such busways, the enclosure has sufficient conductivity. Good electrical connections at the joints also must be checked.

Long conduit runs.

Let us examine a case where a 3-inch conduit run is 1000 feet long. Assume that the minimum fault current would be 600 percent of the overcurrent device rating, or, in this case, 2400 amperes. The impedance of 3-inch conduit at 2400 amperes is approximately 0.0875 ohms/1000 feet. For a 208Y/120 volt system with zero impedance at the point of fault, the maximum current that will flow will be about 1400 amperes. That value of current would operate a 400-ampere high-interrupting capacity current-limiting fuse in about two seconds. However, such ground faults are nearly always arcing faults. This adds resistance to the circuit and reduces the fault current that will flow in the circuit. Moreover, since conduit impedance increases with decrease in fault current, we would have a conduit impedance of 0.129 ohms/1000 feet at a current of 1400 amperes. Allowing for the arc impedance in the circuit described and the increase in conduit impedance at 1400 amps (about 50 percent) and other variable factors including a further increase of conduit impedance at the still lower current, the ground-fault current flow is more likely to be in the order of about 300 amperes. With a 300-ampere fault current, the 400-ampere high-interrupting capacity current-limiting fuse obviously would not clear the fault.

For a long feeder such as this, it may be necessary to increase conductor sizes to account for voltage drop. Section 250-95 requires that the equipment grounding conductor also be increased in size in proportion to size of the ungrounded conductors. For example, if the 3-inch conduit run was only 100 feet long, the anticipated fault current would be about 3,000 amperes, which would satisfactorily operate the overcurrent device. Both fault-current values were based on the impedance of only that part of the circuit that would be within the conduit, not the impedance of the entire circuit. The fault-current values met in practice would likely be less than those given here.

What can we do to get maximum safety? We must design the system to see that grounding fulfills ALL the requirements, as set forth in Section 250-51, to obtain proper or effective grounding, which may mean adding an equipment-grounding conductor within the conduit in parallel with the conduit if the calculations indicate its need.

Where an equipment-grounding conductor is added inside a raceway, we do so to increase the capacity, that is, decrease the impedance of the raceway that is serving as an equipment grounding conductor. Where we have a metallic system, the conductor enclosure also is a part of the equipment-grounding conductor. The assistance of that parallel conduit path is valuable, but proper use can be made of it only if we frequently bond the equipment grounding conductor and the conduit together as frequently as is practical.

If the impedance of the ground-fault circuit is higher than will pass enough current to properly operate the overcurrent devices in a reasonable time, then a lower impedance can be obtained by adding an equipment grounding conductor within the conduit in parallel with the conduit. That conductor must be run **within the conduit,** that is, run with the circuit conductors. The equipment grounding conductor should **NEVER** be run outside the conduit or raceway through which the conductors serving the equipment are run. Where run external to the conduit, it becomes quite ineffective in the grounding circuit, for virtually all the ground-fault current will return on the conduit. Further, the equipment grounding conductor must be run as close to the phase conductors as is practical, right to the point where it attaches to the neutral conductor at the service.

By determining the impedance of the circuit involved when a ground fault occurs, we can get an answer as to whether the metallic enclosure will make an acceptable equipment grounding conductor or whether it will be necessary to supplement the enclosure with an equipment grounding conductor. Where an iron conduit is a part of the electric circuit, as it will be where a ground fault occurs, there will be a large increase in both the resistance and reactance of the circuit, which will vary considerably with the amount of fault current.

Long Conduit Runs

Increase size of conduit or install equipment grounding conductor and bond to raceway as often as practicable. Not more than 100 feet apart.

Figure 11-10

Laboratory tests have shown that where a single-phase current flows in a conductor within an iron conduit, the impedance of the circuit is approximately equal to the impedance of the conduit itself. The size of the conductor within the conduit has relatively little effect on the circuit impedance. Also, despite the fact that there are many parallel paths external to the conduit, the current flowing in all the parallel paths will be very small, and under normal conditions would be under 10 percent of the total fault current.

Two other factors are to be taken into account in estimating the ground-fault current flow. They are the effects of the conduit couplings in increasing the impedance of the circuit and the voltage drop across the point of fault. If conduit couplings are installed wrenchtight, as required by Code, the increase in impedance of the conduit with couplings is about 50 percent more than the impedance for a straight run without couplings. This is where the value and importance of installing a properly-sized equipment grounding conductor inside the conduit and bonding it to the conduit at frequent intervals is proven for ensuring low impedance and safety.

Impedance values for conduit can be obtained from manufacturers. By using these impedance values, adjusting for the couplings and allowing 50 volts drop across the fault, we can arrive at a reasonable value for the amount of fault current that will flow in the circuit.

Assume a 200-foot run of 3-inch conduit with 500 kcmil conductors on a 208Y/120-volt circuit protected by 400 ampere overcurrent devices. On a ground fault, the current that will flow will thus be

$$\frac{E \text{ (voltage)}}{Z \text{ (impedance)}}$$

With a 50-volt drop at the fault and Z (impedance) = 0.02970 the current that will flow will be about 2,350 amperes. The use of a 3-inch conduit as the equipment grounding conductor where 400 ampere overcurrent devices are used is, thus, satisfactory for this run.

A simpler method of determining if the conduit or metallic enclosure will perform satisfactorily is to first calculate the minimum desired fault current flow (5 times the overcurrent device rating) which, in this case, is 5 x 400 or 2,000 amperes. Then, on the basis of 70 volts available for a 120-volt-to-ground circuit, calculate Z (impedance), which will be found to be 0.035. A straight run of 3-inch conduit has about 0.099 ohms impedance per M feet at 2000 amperes. To that figure add 50 percent to include a factor of safety. That will give us an impedance of 0.01485 ohms per 100 feet. The impedance value that will allow 2,000 amperes to flow in that circuit was found to be 0.035. Since 235 feet of 3-inch conduit carrying 2,000 amperes will have an impedance of 0.035, we have determined that to have a minimum current flow of 2,000 amperes in a ground fault, we can install up to 235 feet of 3-inch conduit where a 400-ampere overcurrent device is involved.

If 4-inch conduit was used instead of 3-inch, and the overcurrent device rating did not change but remained at 400 amperes, we can again determine the maximum length of conduit by reference to the table. With some interpolation and using the same calculations, we find that 260 feet of 4-inch conduit could be installed and provide a satisfactory equipment grounding conductor for this circuit.

For any circuit and any size conduit, we can thus derive, for any size overcurrent device, the maximum safe length of conduit which will allow a fault current to pass which will be sufficient to facilitate the operation of the overcurrent device. Should the circuit length exceed the maximum safe length as calculated, then it will be necessary to add a metallic (copper or aluminum) equipment grounding conductor in parallel with the conduit or increase the size of the conduit. The equipment grounding conductor size is per Table 250-95.

It would be neither practical nor desirable to substitute a copper or aluminum conductor for the conduit. Rather, add the copper or aluminum conductor, and connect it in parallel with the conduit to form an equipment grounding conductor of two conductors in parallel. The copper or aluminum conductor and the conduit should be connected together at convenient practical intervals, about every 100 feet or less. That will reduce the length of circuit through which the ground-fault current may flow on the conduit alone.

The Code permits an equipment grounding conductor to be bare or insulated. However, if the conductor is bare, there may be some arcing between the bare conductor and the interior of the conduit at points other than the point at which the ground fault occurs. This is due to slight differences in impedance between the raceway and wire, resulting in potential differences which cause the arcing. Such arcing may damage the phase conductors without adding to the proper functioning of the ground-fault circuit. That makes a strong case for the use of insulated equipment grounding conductors where installed in a metallic enclosure.

If aluminum conduit was used instead of steel conduit for the same conditions cited above, (500 kcmil copper conductors, 3-inch conduit and a 400-ampere overcurrent device) the circuit run could be about 900 feet long, and the aluminum conduit would provide a satisfactory equipment grounding conductor. A 3-inch aluminum conduit has a dc resistance of about 0.0088 ohms/1000 feet and 500 kcmil copper cable has a dc resistance of 0.0222 ohms/1000 feet.

Flexible metal conduit is suitable as an equipment grounding conductor for not more than 6-foot lengths in the ground-fault return path and with not more than 20 ampere overcurrent protection of the contained conductors. As such, flexible metal conduit should include a metallic equipment grounding conductor. The Code requires that the various metal raceways shall be so constructed that adequate electrical and mechanical continuity of the complete system be secured. However, owing to the various joints involved, it is well for the engineer to investigate such conductor enclosures to be assured that their impedance is sufficiently low to function along the lines cited above in the event of a ground fault.

Adequate size of conductor.

In general, the minimum size of equipment grounding conductor is given in Table 250-95 and is based on the rating of the overcurrent device ahead of the conductors. (An analysis of Table 250-95 is provided in Table Seven of Chapter 18.) The rule of thumb may be applied, but it should be checked by calculation that the equipment grounding conductor should not be less than 25 percent of the capacity of the phase conductors or the overcurrent device that supply the circuit. A note has been added to the table indicating that the size of equipment grounding conductor given in the table may need to be increased to comply with Section 250-51. This adds emphasis to the heading of Table 250-95 which indicates that equipment grounding conductors given in the table are the "minimum size."

An analysis of Table 250-95 of the Code shows that the relation of the equipment grounding conductor to the size of the overcurrent device (based on the continuous rating of 75° C-rated wire) is from 125 percent down to 50 percent of the phase conductor for overcurrent devices up to 100-ampere rating. The rating varies from 33 to 25 percent for overcurrent devices rated up to 400 amperes and is from 22 percent for 600 ampere overcurrent devices to a low of only 8 percent for an overcurrent device of 6,000 amperes.

Obviously, the equipment grounding conductor must be large enough to carry that amount of current for a given time, which is required to clear the overcurrent device with which it is associated, in a reasonable time and not result in extensive damage. In addition, it is vital that the equipment grounding conductor be run with the circuit conductors or enclose the circuit conductors. This was covered extensively in Chapter Nine of this text.

Conductor withstand rating.

Section 110-9 of the NEC® tells us that, "equipment intended to break current at fault levels shall have an interrupting rating sufficient for the nominal circuit voltage and the current which is available at the line terminals of the equipment."* This includes fuses, circuit breakers, disconnect switches, contactors, and similar equipment.

Section 110-10 tells us that, "The overcurrent protective devices, the total impedance, the component short-circuit withstand ratings, and other characteristics of the circuit to be protected shall be selected and coordinated to permit the circuit protective devices used to clear a fault to do so without extensive damage to the electrical compo-

nents of the circuit. This fault shall be assumed to be either between two or more of the circuit conductors, or between any circuit conductor and the grounding conductor or enclosing metal raceway."*

Grounding conductors, circuit conductors, bus bars, bonding jumpers, etc., are NOT intended to break current. These conductors must be large enough to safely carry any short-circuit and/or ground-fault current for the time that it takes the overcurrent protective device to clear the fault. This is clearly stated in Sections 110-10, 240-1 Fine Print Note (FPN), 250-1 (FPN No. 1 and 2), 250-51, 250-70, and 250-75. Section 310-10 also discusses in detail the temperature limitations for conductors.

The integrity of equipment grounding conductors, grounding electrode conductors, main bonding jumpers and other circuit conductors must be assured by sizing them properly.

**Wire, Insulation and Connection
Withstand Ratings**

Copper conductor five-second melting rating.
See Column 8, Fig. 11-14

Conductor insulation five-second withstand rating.
See Column 4, Fig. 11-14

Bolted connection five-second withstand rating.
See Column 7, Fig. 11-14

Figure 11-11

Equipment-grounding conductors that are too small are of little value in clearing a fault and may, in fact, give one a false sense of security. Grounding conductors must not burn off under ground-fault conditions, leaving the equipment "live" thus, in many cases, creating a shock hazard which could be fatal.

Safety cannot be compromised. In fact, the first sentence of the National Electrical Code® states that, "the purpose of this Code is the practical safeguarding of persons and property from hazards arising from the use of electricity."*

Bolted connections.

It can be shown by calculation, using values from the Insulated Cable Engineers Association Publication P 23-382, that an insulated copper conductor with a bolted connection may safely carry one ampere for every 42.25 circular mils for five seconds without destroying its validity. That will be the short-time rating or I^2t (amperes x amperes x time) value of the conductor. Then, from the time-current characteristic curves of various approved overcurrent devices, we can learn how much current will be necessary to clear the overcurrent device in five seconds.

Allowable Short-Circuit Currents
For Insuated Copper Conductors

Figure 11-12

Using that formula, we can then calculate the size of equipment grounding conductors which shall be proportional to those given in Table 250-95, as analyzed in Table Seven of Chapter 18.

Grounding conductors are permitted to be bare, and, in many cases, are pulled into the same raceway as the insulated phase conductors. This presents a potential problem. Where a grounding conductor is carrying ground-fault current, an extreme rise in its

temperature can cause the insulation on the adjacent phase conductors to melt, causing further damage. Again, we have increased the potential for equipment damage and electrical shock hazard to personnel.

Thus, as discussed in this chapter, for copper conductors, we must limit the clearing time and short-circuit current to:

One ampere...

for five seconds...

for every 42.25 circular mils.

This can be expressed by the term "ampere squared seconds (I^2t)."

For example, from Table 8 of Chapter 9 of the NEC®, a No. 8 AWG conductor has a cross-sectional area of 16,510 circular mils. By dividing the circular mil area of the conductor by 42.25, we can calculate the conductor's five-second withstand rating.

$$\frac{16,510}{42.25} = 391 \text{ amperes}$$

Stated another way, this conductor has an I^2t five-second withstand rating of: 391 x 391 x 5 = 764,405 ampere squared seconds.

From this five-second withstand rating value, it is easy to calculate the conductor's withstand rating for other values of time and/or for other values of current.

Example 1: How many amperes will the No. 8 AWG copper conductor be able to safely carry if the impedance of the circuit along with the operating characteristics of the overcurrent device protecting the circuit results in a 2 cycle (0.0333 seconds) opening time? Example 1 is solved as follows.

$$I^2t = 764,405 \text{ ampere squared seconds}$$
$$I^2 = \frac{764,405}{t}$$
$$I = \sqrt{\frac{764,405}{0.0333}}$$
$$I = 4,791 \text{ amperes}$$

Example 2: How many amperes will the No. 8 AWG copper conductor be able to safely carry if the impedance of the circuit along with the operating characteristics of the overcurrent device protecting the circuit results in a ¼ cycle (0.0042) opening time? Example 2 is solved as follows:

$$I^2t = 764,405 \text{ ampere squared seconds}$$
$$I^2 = \frac{764,405}{t}$$
$$I = \sqrt{\frac{764,405}{0.0042}}$$
$$I = 13,491 \text{ amperes}$$

Note that in the example above, because we have used a much faster total clearing time, the allowable fault current that the conductor will be subjected to can be increased. This is a result of substituting different time values in the I²t formula.

Generally, where current-limiting overcurrent devices are protecting the circuit, the conductor sizes are determined directly from Table 250-94 and 250-95. Where available fault currents are high, and the overcurrent protective device takes longer than ¼ cycle to clear the fault, is it suggested that the grounding conductor be sized per Figure 11-12, to be on the safe side.

Figure 11-14 provides information to assist the installer in the proper selection of equipment grounding conductors. Among other information, it includes:

1. safe values for 75° C thermoplastic insulated conductors,
2. safe values for bolted connections,
3. unsafe (melting) values for the copper conductor itself.

Because the "weakest link" in any system is the insulation short-circuit withstand rating as found in columns 4, 5, and 6 of Figure 11-14, it is recommended that column 4 be the deciding factor where selecting equipment grounding conductors. The Insulated Cable Engineers Association data, Figure 11-14, Column 4 calculates out to the previously discussed:

Do not exceed one ampere...

for five seconds...

for every 42.25 circular mils of copper conductor.

This chart also shows the No. 8 AWG conductor used in the above text examples.

Where you are absolutely certain that the equipment grounding conductor will not come into contact with any of the current-carrying insulated circuit conductors, the withstand rating of the grounding conductor can be established as follows:

Do Not Exceed One Ampere...

for five seconds...

for every 29.1 circular mils of copper conductor. See Column 7, Figure 11-14.

This is the standard previously referred to in this text as the "Do not exceed one ampere for five seconds for every thirty circular mills of conductor area." This value is only to be used where bare equipment grounding conductors are used in such a manner that they will not come into contact with insulated conductors. In this application, the limiting element of the circuit is the bolted connection of the lug.

Typical Short-Circuit Let-Through Chart

Figure 11-13

Column 8 of Figure 11-14 gives the current in amperes at which the melting temperature of copper conductors is reached. Of course, you never want to reach the current shown as the equipment grounding conductor will burn off leaving the equipment ungrounded and a possible shock hazard.

Conclusion.

That brings us to the inevitable conclusion that in any grounded system, in addition to a good grounding electrode with as low a resistance as practical, it is vital to safety that we have a low-impedance equipment grounding conductor or path. This allows for sufficient current to flow to clear a ground fault automatically in a limited time, which would be as quickly as is practical, without undue interruption of service.

The I²t values found in Column 7 of Figure 11-14 are based on the adequacy of a copper conductor and its bolted joints to carry the current values without destroying its validity. The values are obtained from an IEEE committee report in "A Guide to Safety in AC Substation Grounding." The figure expressed in amperes per circular mil is one ampere for every 29.1 circular mils cross section. The time of five seconds was used to provide a safety factor and was considered a reasonable approach for distribution systems of 600 volts or less protected by high-

Copper 75° C. Thermoplastic Insulated Conductor, Bare Conductor, Bolted Connection Five Second Withstand Rating In Amperes and Melting of Copper Wire

1	2	3	4	5	6	7	8
Wire size	Area in circular mils	Area in square mm	ICEA Amperes	IEC Amperes	IEE Amperes	Bolted connection 250°C. Amperes	Melting of conductor 1,083°C. Amperes
14	4,110	2.080	97	107	107	141	254
12	6,530	3.310	155	170	170	224	403
10	10,380	5.261	246	271	271	357	641
8	16,510	8.367	391	430	430	567	1,020
6	26,240	13.300	621	684	684	902	1,621
4	41,740	21.150	988	1,088	1,088	1,435	2,578
3	52,620	26.670	1,245	1,372	1,372	1,808	3,251
2	66,360	33.620	1,571	1,729	1,729	2,281	4,099
1	83 690	42.410	1,981	2,181	2,181	2,876	5,170
1/0	105 600	53.490	2,499	2,751	2,751	3 629	6,523
2/0	133,100	67.430	3,150	3,468	3,468	4,574	8,222
3/0	167,800	85.010	3,972	4,372	4,372	5,767	10,366
4/0	211,600	107.200	5,009	5,513	5,513	7,272	13,071
250	250,000	126.700	5,918	6,516	6,516	8,592	15,443
300	300,000	152.000	7,101	7,818	7,818	10,310	18,532
350	350,000	177.300	8,285	9,119	9,119	12,029	21,621
400	400,000	202.700	9,467	10,425	10,425	13,747	24,709
500	500,000	253.300	11,834	13,027	13,027	17,184	30,887
600	600,000	304.000	14,201	15,636	15,636	20,621	37,064
700	700,000	354.700	16,568	18,243	18,243	24,057	43,241
750	750,000	380.000	17,752	19,544	19,544	25,776	46,330
800	800,000	405.400	18,935	20,850	20,850	27,494	49,419
900	900,000	456.000	21,302	23,453	23,453	30,931	55,596
1,000	1,000,000	506.700	23,669	26,060	26,060	34,368	61,773

Column 4 - Insulated Cable Engineers Association publication P32-382. One ampere for five seconds for every 42.25 circular mils of conductor area.

Column 5 - International Electrotechnical Commission publication 364-4-43.

Column 6 - Institute of Electrical Engineers publication 434-6.

Column 7 - Calculated from data in Electrical Engineers Handbook (75°C ambient). One ampere for five seconds for every 29.1 circular mils of conductor area.

Column 8 - Calculated from data in Electrical Engineers Handbook (75°C ambient). One ampere for five seconds for every 16.19 circular mils of conductor area.

Figure 11-14

interrupting-capacity current-limiting fuses and having equipment ground-fault protection.

As previously stated, where grounding conductors might come into contact with insulated phase conductors, use the values found in Column 4 of Figure 11-14. This column is based on one ampere for every 42.25 circular mils of conductor for five seconds. Figure 11-14 has all the calculations done and is much easier to use rather than performing complicated calculations.

The short-time rating of the equipment grounding conductor bears an approximately constant relation to the size of the overcurrent device. The I^2t values of the conductors given in Table 250-95 are between about 13 and 28 times their nominal continuous rating based on one ampere for every 42.25 circular mils cross section.

References

[1] *Electric Power Distribution for Industrial Plants. The Institute of Electrical and Electronics Engineers, Inc., © 1954 AIEE (now IEEE), 445 Hoes Lane, P.O. Box 1331, Piscataway, NJ 08855-1331.*

[2] *Portions of "Some Fundamentals of Equipment-Grounding Circuit Design" by R. H. Kaufmann, Paper 54-244 presented at the AIEE Summer and Pacific General Meeting, Los Angeles, California, June 24-25, 1954, © 1954 AIEE, (now IEEE), 445 Hoes Lane, P.O. Box 1331, Piscataway, NJ 08855-1331.*

[3] *1991 IEEE Industry Applications Society Annual Meeting, Volume II. The Institute of Electrical and Electronics Engineers, Inc., © 1954 AIEE (now IEEE), 445 Hoes Lane, P.O. Box 1331, Piscataway, NJ 08855-1331.*

[4] *Modeling and Testing of Steel EMT, IMC and Rigid (GRC) Conduit School of Electrical and Computer Engineering*
Georgia Institute of Technology
Atlanta, Georgia 30332

Chapter Eleven: The questions included here were developed using material included in this chapter. The answers can be found by reviewing the text. It is also important that students make use of the 1996 NEC®, where many answers can be found. See page 237 for answers.

1. A conducting connection, whether intentional or accidental, between any of the conductors of an electrical system whether it be from line to line or line to the grounded conductor, is defined as a ____.
   a. ground fault
   b. phase fault
   c. short circuit
   d. unidentified fault

2. A conducting connection, whether intentional or accidental, between any of the conductors of an electrical system and the conducting material which encloses the conductors, or any conducting material that is grounded, or that may become grounded is defined as a ____.
   a. identified fault
   b. ground fault
   c. short circuit
   d. failure of the system

3. A short circuit may be from one phase conductor to another phase conductor, or from one phase conductor to the grounded conductor or ____.
   a. unidentified conductor
   b. enclosure
   c. neutral
   d. equipment grounding conductor

4. The grounding system must provide a permanent and continuous path; must have ample carrying capacity to conduct safely any fault currents likely to be imposed on it; and must have an impedance sufficiently low to limit the voltage to ground and to facilitate the operation of the circuit protective devices in the circuit. This is best described as being ____.
   a. improved
   b. effective
   c. sufficient
   d. required

5. In general, the rule of thumb may be applied, but should be checked by calculation, that the equipment grounding conductor should not be less than ____ percent of the capacity of the phase conductors that supply the circuit.
   a. 25
   b. 20
   c. 15
   d. 18

6. There is no overcurrent protection on the line side of service conductors, there is only ____ protection on the primary of the transformer.
   a. ground-fault
   b. short-circuit
   c. overload
   d. GFPE

7. The grounded service conductor must be run to each service disconnecting means and bonded to the enclosure regardless of whether or not the service supplies only ____.
   a. 240 volt loads
   b. 208 volt loads
   c. 480 volt loads
   d. line-to-line loads

8. A No. 8 THW copper conductor has an allowable ampacity of 50 amperes according to Table 310-16. This conductor also has a safe (ICEA) five-second withstand rating of ____ amperes, but will melt if subjected to ____ amperes for five seconds. See Figure 11-14 in text.
   a. 8,367   10,000
   b. 391   1,020
   c. 30   500
   d. 4,110   254

9. When ICEA conductor short-circuit withstand rating tables are not available, it is possible to calculate the short-circuit current withstand ratings for a conductor. For example, to determine the safe withstand rating for copper conductors with 75° C insulation, use ____ ampere for every ____ seconds for every ____ circular mils of cross sectional area of the conductor.
   a. one, five, 42.25
   b. one, five, 16.9
   c. one, five, 29.1
   d. one, five, 10.2

10. Table 250-95 shows the minimum size equipment grounding conductors. Where high values of fault current are available, the grounding conductors may have to be ____ in size to be capable of safely carrying the available fault current.
   a. decreased
   b. increased

11. Referring to Figure 11-12, find the minimum conductor sizes required for the fault current values indicated.
   a. 10,000 amperes for one cycle.   Size ____
   b. 20,000 amperes for two cycles.   Size ____
   c. 4,200 amperes for one cycle.   Size ____
   d. 10,000 amperes for eight cycles. Size ____

Chapter 12

Grounding Separately Derived Systems

Objectives

After studying this chapter, the reader will be able to understand

- General requirements and definitions for separately derived systems.
- Installation and sizing of bonding jumper for separately derived systems.
- Sizing and types of grounding electrode conductors for separately derived systems.
- Transformer overcurrent protection.
- Generator types and systems.
- Ground-fault systems.

Definition.

Separately Derived System. "A premises wiring system whose power is derived from a battery, a solar photovoltaic system, or from a generator, transformer, or converter windings, and that has no direct electrical connection, including a solidly-connected grounded circuit conductor, to supply conductors originating in another system."* (Article 100.)

General.

In many distribution systems of 600 volts or less, we are liable to have separately derived systems at another voltage level lower or higher than the originating system. By reference to Section 250-5, one can determine whether or not a separately derived system must be grounded. Where the derived system meets the criteria found in Section 250-5, it must be grounded according to the requirements of Section 250-26. Where the derived system is not required to be grounded by Section 250-5, grounding of the derived system is optional.

Each voltage level that is required to be grounded must be grounded independently at its source. The equipment grounding conductor of each voltage level must meet the requirements of Section 250-51, in being electrically continuous and permanent, of ample capacity to carry the currents that may be expected and be connected to the grounded conductor (at that voltage level) within the enclosure, as is done in the case of service equipment. That connection will become a common point for the grounded conductor and the equipment grounding conductor where the common conductor may be attached and run to the grounding electrode.

Transformer type.

Figure 12-1 is a one-line diagram of a transformer type separately derived system. (For simplicity, grounding and bonding conductors are not shown.) The supply to the transformer is from one voltage level, and the secondary is often at another voltage level, either lower or higher. Where 3-phase systems are employed, the primary is usually connected delta. The secondary may be connected delta or wye as needed.

Grounding primary side equipment.

An equipment grounding conductor must be supplied with the primary circuit to provide a low-impedance fault-current path from the transformer case to the main service. The overcurrent device on the primary of the transformer will then clear a

Transformer-Type Separately Derived System

A premises wiring system whose power is derived from a battery, a solar photovoltaic system, or from a generator, transformer, or converter windings, and that has no direct electrical connection, including a solidly connected grounded circuit conductor, to supply conductors originating in another system.

Figure 12-1

short circuit or ground fault up to, and including, the transformer primary windings. A short circuit or ground fault on the secondary of the transformer is seen as a load by the transformer primary winding. The amount of current flowing in the ground fault on the secondary side will determine whether or not the primary overcurrent device will open.

Bonding jumper.

Section 250-26(a) requires a bonding jumper be installed for the derived system that is required to be grounded by Section 250-5.

Section 250-26(a) permits the bonding jumper to be installed at any point from the source of the separately derived system to the first system disconnecting means or overcurrent device. To avoid confusion, it is best to install the bonding jumper and make the grounding electrode connection at the same point. Many transformer manufacturers supply a bonding jumper and terminals for connection inside the transformer enclosure.

In Figure 12-2, the neutral of the separately derived system is shown connected inside the transformer case to the case through the bonding jumper. The bonding jumper is required to be sized according to Table 250-94 based on the size of the derived phase conductors on the secondary side of the transformer. For example, if the derived phase

Bonding Jumper Location

Bonding jumper
in transformer

Bonding jumper
connection permitted at
any point from source to
disconnecting means or
overcurrent device.
Make at source that has
no disconnecting means
or overcurrent device.

Bonding jumper
in panelboard

Figure 12-2

conductors are 3/0 copper, the bonding jumper must be no smaller that No. 6 copper or No. 4 aluminum. In this regard, requirements for separately derived systems are similar to services. Like services, the size of the bonding jumper is not based on the rating of the main overcurrent device or on the rating of the equipment being supplied but on the size of the ungrounded conductor supplied.

An alternate, and equally correct, method would be to bond the derived neutral to the neutral terminal bar in the distribution cabinet. As indicated in Section 250-26(a), the derived neutral could be bonded at any other convenient point between the source of power and the first system disconnecting means or overcurrent device such as a pull or junction box.

The bonding jumper is a most critical part of the

separately derived system. Without it, there will not be a path for ground faults on the derived system to return to the source. All ground-fault current downstream from the derived system will pass over the bonding jumper as it returns to its source. This ground-fault current from the derived system will not attempt to return to the service equipment as that is not its source.

Bonding jumper exception.

Section 250-26(a) recognizes the provision for connecting the main bonding jumper according to Section 250-23(a) Exception No. 4 or 5.

Exception No. 4 of Section 250-23(a) permits the grounding electrode connection to be made at a common tie point of dual fed (double ended) services. By making this reference, the implication is that the bonding jumper of a separately derived system is permitted to be made at a similar location.

Bonding Jumper Exception

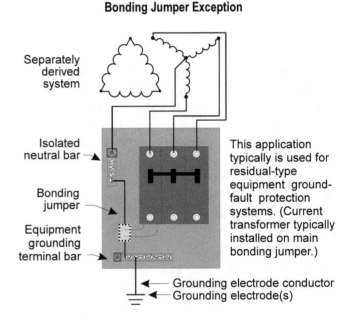

Separately
derived
system

Isolated
neutral bar

Bonding
jumper

Equipment
grounding
terminal bar

This application
typically is used for
residual-type
equipment ground-
fault protection
systems. (Current
transformer typically
installed on main
bonding jumper.)

Grounding electrode conductor
Grounding electrode(s)

Figure 12-3

Exception No. 5 of Section 250-23(a) permits the grounding electrode connection to be made to an equipment grounding terminal bar, rather than to the neutral bar where the main bonding jumper is a wire or busbar, as permitted by Sections 250-53(b) and 250-79.

One application of this exception is for residual-type equipment ground-fault protection systems where, often, a current transformer is located on the main bonding jumper to measure ground-fault currents flowing back to the source. Another common application of this exception is in switchboards where the main bonding jumper is a wire or busbar.

Use of derived neutral for bonding.

Section 250-61(a)(3) permits the grounded circuit conductor to be used for grounding equipment "On the supply side of the disconnecting means or overcurrent devices of a separately derived system." As used in Section 250-61(a), the term, "on the supply side of" no doubt means "up to or within the enclosure for the disconnecting means or overcurrent devices" as it is widely accepted to make bonding connections of the neutral or grounded circuit conductor within the disconnecting means enclosure for the service, building and separately derived system.

Use of Derived Neutral for Bonding

Wiring method does not serve as equipment ground

Figure 12-4

As shown in Figure 12-4, the grounded circuit conductor serves two purposes. First, it allows line-to-neutral or grounded conductor loads to be supplied, and the conductor carries the unbalanced loads from ungrounded conductors. Secondly, it serves as the equipment grounding conductor and will carry line-to-ground fault currents back to the source. This grounding scheme parallels that for services and for supply to additional buildings on the premises served from a common service.

Where this scheme is used, it is not necessary to install an equipment grounding conductor between the source and the disconnecting means or overcurrent protection enclosure. This will allow nonmetallic raceways to be used. Where nonmetallic raceways are used, a parallel path for neutral current does not exist.

A parallel path for neutral current exists where a metal raceway is used between the source of the separately derived system and the metal enclosure for the disconnecting means or overcurrent device. The neutral current will divide between the available paths depending upon the impedance of the paths. The lowest impedance path will obviously carry the most current. The parallel path also exists for neutral current where a metal raceway is used for services and second or additional buildings and the neutral is bonded to the metal enclosures at each end of the conductor.

Care should be taken when using this method with flexible metal conduit or liquidtight flexible metal conduit. Flexible metal conduit is limited to a maximum of 20 amperes where used as an equipment grounding conductor by Sections 250-91(b) Exception No. 1 and 350-14 Exception. Liquidtight flexible metal conduit is limited as an equipment grounding conductor to 20 amperes in the 3/8- and ½-inch sizes and 60 amperes in the ¾-inch through 1¼-inch sizes. Undesirable heat can be created by current flow in these wiring methods where a parallel path for neutral current is established by use of the bonded grounded conductor and one of these metallic raceways.

Grounding electrode conductor.

Next, a grounding electrode conductor must be installed to comply with Section 250-26(b). The size of this conductor is determined by Table 250-94, based upon the size of the derived phase conductors. For example, if the derived phase conductors are 500 kcmil aluminum, the minimum size grounding electrode conductor determined by reference to Table 250-94 is No. 2 copper or 1/0 aluminum.

Again, the connection of the grounding electrode conductor is permitted to be made at the transformers, the first disconnecting means or overcurrent device, or any point between. To avoid confusion, it is best to make the grounding electrode connection to the same point the bonding jumper is installed. Exceptions are provided for Class 1 transformer derived circuits (usually control circuits) and for high-impedance grounded neutral systems.

Grounding Electrode Conductor Connections

Grounding electrode connection in transformer

Grounding electrode connection permitted at any point from source to disconnecting means or overcurrent device. Make at source that has no disconnecting means or overcurrent device.

Grounding electrode connection in panelboard

Figure 12-5

Grounding electrode.

The Code, in Section 250-26(c), is very specific about the grounding electrode required to be used for separately derived systems. The grounding electrode is to be:

(1) as near as practicable to, and preferably in the same area as, the separately derived system.

(2) the nearest available effectively grounded structural metal member of the structure; or

(3) the nearest available effectively grounded metal water pipe; or

(4) other electrodes as specified in Sections 250-81 and 250-83 where electrodes indicated in (2) or (3) are not available.

> ## Grounding electrode for separately derived systems.
>
> - **Near as practicable.**
> - **Nearest available effectively grounded structural metal member.**
> - **Or nearest available effectively grounded water pipe.**
> - **Other electrodes where the above are not available.**

The Code goes on to say that in all other respects, grounding methods shall comply with other requirements in the Code. This includes requirements for installation methods and protection of conductors from damage and connection of the grounding electrode conductor to grounding electrodes.

Section 250-54 requires that the same electrode be used for grounding all ac systems in or at the building. Where more than one electrode is installed, they then must be bonded together. This applies to services as covered in Section 250-23 and to two or more buildings or structures in Section 250-24, but does not directly apply to separately derived systems. However, from an electrical perspective, there is no reason why all the grounding electrodes should not be bonded together. Since the building service is grounded to effectively grounded building structural steel, and the interior water pipe is required to be bonded, the separately derived system and the service neutral will be electrically connected together.

In some installations, the same grounding electrode will be used for both the main or utility supplied electrical system and the separately derived system. This occurs where an effectively grounded structural metal member of the structure or effectively grounded water pipe, as directed in Section 250-81, is available and used for the grounding electrode. In other cases where these "system" grounding electrodes are not available, other electrodes such as concrete encased electrodes, ground rings or made electrodes, such as metal pipes, tanks, rods and plates, are required to be used.

Locating the grounding electrode(s) in the same area as the system helps avoid shock hazards that can result from electrical systems having a ground potential different from the potential where the system is located.

Ground-fault return path.

As with all electrical systems, it is vital to be certain that, for separately derived systems, an effective ground-fault return path exists from the farthermost point on the electrical system back to the source.

Fault current downstream from the separately derived system follows the same laws of physics as does the fault current on the line side. It attempts to return to its source, which is the secondary windings of the transformer, and seeks the lowest impedance path. As can be seen in the figure, fault current downstream from the separately derived system will return to the secondary windings of the transformer, not to the primary or to the service. The

transformer primary sees the fault current on the secondary as a load, and the overcurrent devices on the line side of the transformer will respond according to their rating.

As illustrated in Figure 12-6, be certain that a path for ground-fault current is permanent and continuous, of adequate capacity, and low impedance from the equipment supplied back to the separately derived system. Often, it is best to make a one-line diagram of the installed system to be certain a complete and low-impedance path is provided.

One fairly common error in installing these systems is to use flexible metal conduit between the transformer and panelboard without installing an equipment grounding conductor through the flexible conduit. As fully explored in Chapter 5 of this textbook, flexible conduit is suitable as an equipment grounding conductor only where the contained conductors have overcurrent on the line side not greater than 20 amperes. This requirement practically prevents flexible metal conduit from being used between the transformer and the panelboard or fused switch unless an equipment grounding means is installed between the transformer and the overcurrent device.

If the bonding jumper is installed in the panelboard or fusible switch rather than in the transformer, and flexible metal conduit is used as the wiring method, an equipment grounding conductor must be installed between the transformer enclosure and the overcurrent device enclosure. In this case, a parallel path will exist through the equipment grounding conductor and the metal wiring method.

Ground-Fault Return Path

Equipment grounding conductors.

The equipment grounding conductors, of both the primary and the separately derived system, may be obtained by using the metal conductor enclosures, such as conduit or electrical metallic tubing, where permitted by Code. An equipment grounding conductor may be used, if necessary, to be certain the circuit is of low enough impedance to be effective. In both the primary and the separately derived system, a low-impedance path to the neutral of each system respectively must be provided. This is vital to ensure that overcurrent devices will function to remove ground-faults that may occur.

The important connections here require an equipment grounding conductor for both the primary and separately derived systems independent of each

Figure 12-6

other and terminating at the respective points of supply at their grounded conductor or neutral. Both equipment grounding conductors will meet at some common point, as at the point where the bonding jumper is connected, and they both may connect to the common grounding electrode.

Interconnecting the equipment grounding conductor of the two systems does not violate the definition of separately derived system. This is due to the fact that the equipment grounding conductor is not a power conductor nor a grounded conductor. Also, the equipment grounding conductor usually does not make a direct electrical connection of the neutral to ground. By following, on a one-line diagram, the bonding and equipment grounding conductors of the two systems, it can be seen that the systems are, in fact, connected together by these conductors.

Bonding metal water piping.

Special bonding rules apply where a separately derived system is installed in an area where there is no effectively grounded structural building steel nor effectively grounded metal water piping. In this case, the grounding electrode is permitted to be a concrete encased electrode, a ground ring or a made electrode such as a ground rod from Section 250-83.

Water pipe in the area served by the separately derived system must be bonded to the derived grounded conductor at the nearest available point. This will ensure that this metal water piping is not

electrically isolated and, should it become energized, become a shock hazard.

The bonding conductor must be sized in accordance with Table 250-94 based on the size of the derived phase conductors and be installed in compliance with Sections 250-92(a) and (b).

Transformer overcurrent protection.

Overcurrent protection of the primary as well as the secondary disconnect for the transformer must comply with Section 450-3. In addition, Section 384-16(d) requires that where a transformer supplies a panelboard, overcurrent protection, not exceeding the rating of the panelboard, be provided on the secondary of the transformer.

An exception, excluding the requirement for the supply conductors terminating in an overcurrent protective device in a panelboard, is provided for 2-wire to 2-wire, single-voltage transformers.

Generator-type separately derived system.

Figure 12-8 illustrates a separately derived system from a generator. Note that there is no connection, including that of a solidly grounded neutral, between the two systems. An easy way to determine whether or not the generator is a separately derived system is to examine the transfer switch. If the neutral and all phase conductors are switched, it is a separately derived system. If the neutral is not switched, but is solidly connected, then it is not a separately derived system.

Where the generator is a separately derived system, the neutral bonding jumper must be installed, either at the generator or at the first disconnecting means or overcurrent device or any point

Bonding Metal Water Piping

Separately derived system

Bond interior metal water pipe at the nearest available point in the area served by the separately derived system, to the grounded conductor -

where the system is not grounded to an effectively grounded structural metal member or effectively grounded metal water pipe.

Figure 12-7

Generator-Type Separately Derived System

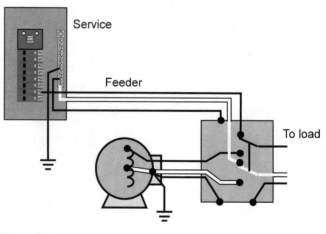

Service

Feeder

To load

Figure 12-8

Signs for Emergency Sources

Figure 12-9

**Generator Power Supply
Not Separately Derived System**

Figure 12-10

between. In addition, a grounding electrode conductor must be installed between the neutral and a grounding electrode.

Section 700-8(b), regarding grounding of emergency systems, has been added to require that where the grounded circuit conductor connected to the emergency source is connected to a grounding electrode conductor at a location remote from the emergency source, a sign must be placed at the grounding location that shall identify all emergency and normal sources connected at that location.

Identical changes were made to Section 701-9(b) for legally required standby systems and to Section 702-8(b) for optional standby systems.

Alternate power supply systems that have an unswitched neutral, that is grounded at the service entrance, present a safety concern, as electricians that are unfamiliar with the system grounding scheme may inadvertently disconnect the grounded (neutral) conductor when working on the normal source. If the emergency system is operating, the neutral conductor from the transfer switch to the location where it is grounded will function as a grounding electrode conductor, and disconnecting it may pose a serious shock hazard.

Generator not separately derived.

Where the generator is not a separately derived system, the neutral bonding jumper must be removed from the generator, and the neutral must not be grounded at the generator or at any point between the generator up to the service.

In this case, the system is grounded by its solid connection to the neutral of the premises wiring system. An equipment grounding conductor is installed throughout the system with the circuit conductors between the service, transfer switch, and other noncurrent carrying parts of the installation.

Ground-fault systems.

A word of caution. It is imperative that care be exercised where generators supply systems that have equipment ground-fault protection. In most cases, the engineer will specify that the neutral be switched by the transfer switch to avoid grounding the neutral downstream from the ground-fault protection equipment. This is vital to prevent desensitizing the protection system. In this case, the generator is a separately derived system, and it must be grounded and bonded separately.

Section 700-26 indicates that equipment ground-fault protection with automatic disconnecting means is not required for emergency systems. However, indication of a ground fault on the emergency system must be given in accordance with the rules of Section 700-7(d). There, it is required that, where practicable, audible and visual signal devices be provided that will indicate a ground fault in a solidly-grounded wye connected emergency system of more than 150 volts to ground and circuit protective devices rated 1,000 amperes or more. A similar exception from the requirement for ground-fault protection of equipment is provided in Section 701-17 for legally required standby systems.

Chapter Twelve: The questions included here were developed using material included in this chapter. The answers can be found by reviewing the text. It is also important that students make use of the 1996 NEC®, where many answers can be found. See page 237 for answers.

1. A premises wiring system whose power is derived from generator, transformer, or converter windings without any direct electrical connection, including a solidly-connected grounded circuit conductor to supply conductors originating in another system, is defined as a ____.
   a. medium voltage system
   b. individual system
   c. separately derived system
   d. high voltage system

2. Phase conductors used for a separately derived system sized at 500 kcmil aluminum, require a grounding electrode conductor not smaller than No. ____.
   a. 4 copper or 2 aluminum
   b. 2 copper or 1/0 aluminum
   c. 6 copper or 4 aluminum
   d. 8 copper or 6 aluminum

3. The connection of the grounding electrode conductor is permitted to be made at the transformer, or the ____ disconnecting means or overcurrent device or any point between.
   a. first
   b. second
   c. any
   d. line-side

4. The grounding electrode for a separately derived system is required to be located as near as practicable to, and preferably in, the ____ as the separately derived system.
   a. same enclosure
   b. same cabinet
   c. same area
   d. same pullbox

5. A grounding electrode conductor for a separately derived system is required to be connected to the nearest available, effectively grounded ____ in the structure.
   a. underground metal gas pipe
   b. structural metal member
   c. service equipment enclosure
   d. metal raceway

6. Where the grounding electrode for a separately derived system, as in question No. 5, is not available, the grounding electrode is required to be the nearest available, effectively grounded ____
   a. metal enclosure
   b. metallic gas pipe
   c. metal water pipe
   d. metal plate

7. Where installed in or at the building, the same ____ is required to be used for the grounding of all ac systems.
   a. electrode
   b. raceway
   c. enclosure
   d. bonding jumper

8. The ____ grounding conductors of both the primary and the separately derived system are permitted to be obtained by using the conductor enclosures, where permitted by Code, or a ____ grounding conductor is permitted to be used, if necessary, to have a circuit with a low impedance.
   a. isolated
   b. identified
   c. system
   d. equipment

9. If the neutral and all phase conductors are switched, it ____ a separately derived system.
   a. is
   b. is not
   c. may be
   d. cannot be considered as

10. Where a generator is a separately derived system, the ____ bonding jumper must be installed, either at the generator or at the first disconnecting means or overcurrent device.
    a. equipment
    b. neutral
    c. service
    d. No. 10

11. Where the generator is not a separately derived system, the ____ bonding jumper must be removed from the generator and the neutral must not be grounded. In this case, the system is grounded by its solid connection to the neutral of the premises wiring system.
    a. No. 10
    b. equipment
    c. service
    d. neutral

Notes:

Grounding at More Than One Building

Objectives

After studying this chapter, the reader will be able to understand

- General requirements for grounding at more than one building.
- Sizing of grounding conductors.
- Bonding grounding electrodes together.
- Disconnecting means for separate buildings.
- Objectionable currents.
- Requirements for mobile homes, recreational vehicles, and agricultural buildings.

Definition.

"**Grounding Conductor:** A conductor used to connect equipment or the grounded circuit of a wiring system to a grounding electrode or electrodes."* (Article 100)

This conductor applies to fairly limited applications such as connecting the equipment at a second or additional building to the grounding electrode and is not the general use grounding electrode conductor one may think.

General.

Section 250-24 provides requirements for grounding of electrical systems and equipment at buildings or structures supplied from a common service.

Included are rules for grounded systems, ungrounded systems and grounding at buildings where the disconnecting means are located in a separate building or structure on the same premises.

Grounded systems.

The general rule in Section 250-24(a) is that at each building or structure served a grounding electrode meeting the requirements of Article 250 Part H must be connected to the grounded circuit conductor (usually the neutral) and the building or structure disconnecting means. Exception No. 1 states that "A grounding electrode at separate buildings or structures shall not be required where only one branch circuit is supplied and there is no equipment in the building or structure that requires grounding." This exception will not apply to many installations since most buildings or structures that are supplied with electricity have at least one piece of equipment that requires grounding.

Two general methods are provided in Section 250-24(a) of the National Electrical Code® for grounding electrical systems at additional buildings or structures on the premises. For convenience we will consider them as method 1 where the neutral is regrounded and method 2 where the neutral is not regrounded.

Method No. 1 (shown in Figure 13-1)

In this method, both the grounded circuit conductor (usually a neutral) and the noncurrent-

---

Grounding Neutral at Second Building

Feeder may be overhead or underground.

Figure 13-1

carrying equipment are connected to a grounding electrode (system) at the additional building or structure. The number and type of conductors that must be taken from the building or structure where the service is located to the second building or structure is summarized in the following table.

Table 13-1 Equipment Grounding Conductor Not Installed			
System	Un-grounded	Grounded	Equip. Ground.
120V	1	1	0
120/240V	2	1	0
208/120V	3	1	0
480/277V	3	1	0

In this case, the electrical system at the additional building or structure is treated like a service for grounding purposes, although the building or structure is supplied from a service in another building or structure by a feeder or circuit. The grounded circuit conductor is bonded to the disconnecting means enclosure for the building or structure. The grounding electrode connection to the grounded (often a neutral) conductor must be made on the supply side of or within the building or structure disconnecting means.

Both the grounded (usually a neutral) conductor(s) and the equipment grounding conductor(s) are permitted to be connected to the terminal bar where the grounding electrode conductor connects to the grounded conductor.

Note that the grounded conductor from the building or structure where the service is located to the second or additional building or structure must generally be an insulated conductor. See Section 310-2(a). Section 338-3(b) permits a Type SE cable without insulation on the grounded circuit conductor to be used as a feeder to supply other buildings on the same premises.

Size of grounded conductor.

The grounded circuit conductor must be sized no smaller than an equipment grounding conductor from Table 250-95 where an equipment grounding conductor is not run from the service to the additional building or structure. In this case, the grounded circuit conductor serves three purposes: first, to permit line-to-neutral loads to be utilized; second, to carry unbalanced loads back to the source; and, third, as an equipment grounding conductor. As such it must be sized for the calculated load according to Section 220-22 and not smaller than an equipment grounding conductor from Table 250-95.

Insulated Neutral at Second Building

Feeder

Service

Subpanel →

Feeder may be overhead or underground.

Figure 13-2

Method No. 2 (shown in Figure 13-2)

In this method, an equipment grounding conductor is run with the feeder or circuit, with the ungrounded and grounded conductors to the additional building or structure. (Section 250-24(a) Exception No. 2) The grounded (neutral) conductor should not be grounded at the second or additional building or structure but is insulated from the equipment enclosure. This method is similar to or identical to installing a feeder to a feeder panel in the same building or service where the service is located.

The equipment grounding conductor, that is run from the building or structure where the service is located, is connected to a terminal bar on the supply side of or inside the building or structure disconnecting means where it is grounded by connecting the grounding electrode to the equipment grounding terminal bar.

The grounded (often a neutral) conductor, installed as a part of the feeder or branch circuit from the building or structure where the service is located to the second or additional building or structure, must be an insulated conductor. See Section 310-2(a).

	Table 13-2 Equipment Grounding Conductor Installed		
System	Un-grounded	Grounded	Equip. Ground.
120V	1	1	1
120/240V	2	1	1
208/120V	3	1	1
480/277V	3	1	1

By using this method, the grounded (often a neutral) conductor(s) must be connected to the terminal bar for the grounded conductors, and the equipment grounding conductor(s) must be connected to the equipment grounding terminal bar where the grounding (electrode) is connected inside the building or structure disconnecting means enclosure.

Grounding conductor.

A grounding conductor (defined as grounding electrode conductor at services) is used to connect the grounded circuit conductor or equipment grounding conductor and equipment to the grounding electrode system that exists at the additional building or structure. Where no grounding electrodes exist at the additional building or structure, one or more grounding electrodes must be installed.

This connection must be made inside the building or structure disconnecting means to the equipment grounding terminal bar.

The equipment grounding conductor run with the feeder or circuit from the first to the additional building(s) must be sized in accordance with Table 250-95 based upon the ampere rating of the overcurrent protective device ahead of the feeder or circuit.

Size of Grounding Conductor

Feeder

Service

Table 250-95

Table 250-94

Figure 13-3

Sizing grounding conductor.

The grounding conductor (grounding electrode conductor at services) is sized from Table 250-95 based upon the ampere rating of the overcurrent protective device ahead of the feeder or circuit. As can be seen, this conductor may be smaller than that required for a grounding electrode conductor.

Grounding electrode requirements.

Grounding electrodes that exist at the additional building or structure must be bonded together to form a grounding electrode system in compliance with Section 250-81. The grounding electrode system must be connected to the grounded conductor supplying the additional building or structure or to the disconnecting means enclosure, depending upon the method of supply to the building or structure and the grounding method employed.

Where there are no grounding electrodes at the additional building or structure, a grounding electrode that complies with Part H of Article 250 must generally be installed. These electrodes include underground metal water pipes in contact with the earth for 10 feet or more, an effectively grounded metal frame of the building, concrete encased electrodes or a ground ring. Made electrodes that are permitted to be used include other local metal underground systems or structures such as piping systems and underground tanks, rods, pipes, con-

duit and plate electrodes. See Section 250-83. It is perhaps most common to see ground rods to be installed at these remote building or structure disconnecting means. Keep in mind that the grounding electrode installation must comply with all the requirements of Part H of Article 250. This includes the requirement that the ground rod, where used, must be augmented by an additional electrode where the resistance to ground exceeds 25 ohms.

As mentioned earlier, Exception No. 1 to Section 250-24(a) indicates that a grounding electrode at separate buildings or structures is not required where only one branch circuit is supplied, and there is no equipment in the building that requires grounding. Exception No. 2 to this section exempts installations from the requirement that a grounding electrode and connection to the equipment grounding conductor supplied to the additional building or structure be made where the building or structure is supplied by only one branch circuit.

Bonding grounding electrodes together.

Section 250-54 does not require that grounding electrodes at separate buildings be bonded back to the service. A study of the system will show that we do not really have two isolated grounding electrode systems. What we have is a grounding electrode system for the electrical system at the individual buildings which are connected together either by the grounded conductor or by the equipment grounding conductor. This is illustrated by both Figures 13-1 and 13-2.

Ungrounded systems.

Where more than one building or structure is supplied with an ungrounded electrical system, a grounding electrode conductor must connect the electrical equipment enclosure to the grounding electrode that exists at the building or structure. A grounding electrode system must be installed at the building or structure where none exists.

For ungrounded systems, two exceptions from the requirement to install a grounding electrode at the second building or structure are provided.

The first exception excludes buildings or structures supplied by only one branch circuit, and there is no equipment at the building or structure that requires grounding.

The second exception excludes buildings or structures from the requirement of having a grounding electrode connection to the building or structure disconnecting means, provided that:

Ungrounded System

Feeder may be overhead or underground.

Figure 13-4

Disconnecting Means Not Located at Second Building

Feeder may be overhead or underground.

Figure 13-5

(a) An equipment grounding conductor is run, with the circuit conductors to the metal enclosure of the building or structure disconnecting means, for grounding any noncurrent-carrying equipment, interior metal piping systems and building or structural metal members.

(b) There are no existing grounding electrodes at the building or structure that comply with Part H of Article 250.

(c) The building or structure is supplied by only one branch circuit.

Disconnecting means located in separate building or structure on the same premises.

Special rules have been provided for large capacity, multibuilding industrial installations under single management in Section 250-24(c). These occupancies have trained and qualified personnel and have established procedures for safe switching procedures. As a result, the disconnecting means are permitted to be located at other locations on the premises rather than at the building served.

The special rules for grounding the electrical system at these separate buildings or structures are as follows:

1. The grounded circuit conductor cannot be grounded at the additional building or structure.

2. An equipment grounding conductor must be run with the circuit conductors to the additional building or structure. There, it must be connected to noncurrent-carrying electrical equipment, interior metal piping systems and

building or structural metal frames and to existing grounding electrodes. Where no grounding electrodes exist at the building or structure, a grounding electrode in compliance with Part H of Article 250 must be installed where more than one branch circuit is supplied.

3. The equipment grounding conductor run to the building or structure must be bonded to the noncurrent-carrying electrical equipment and to the grounding electrode in a junction box, panelboard or similar enclosure such as a switchboard or motor control center.

Two exceptions similar to those for the grounded and ungrounded systems are provided.

Use of grounded circuit conductor.

Sections 250-24(a) and 250-61(a)(2) specifically permit the grounded circuit conductor (usually the neutral) to be used for grounding equipment on the supply side of the main disconnecting means for separate buildings or structures that are supplied from a common service. This grounding scheme is covered in detail earlier in this chapter. Without a doubt, we end up with multiple grounding connections of the grounded service conductor as a grounding connection is made at the supply transformer, at the main service and again at each building or structure where the grounded circuit conductor or neutral are grounded again. In most cases, this regrounding should not present a problem.

Since regrounding the neutral at the additional buildings or structures has established parallel paths for neutral current to flow, the current will divide inversely proportional to the impedance and flow in each path. The lowest impedance path will obviously carry the most current. The lowest impedance path will always be through the neutral or grounded circuit conductor rather than through grounding electrodes and the earth or over continuous metal piping that is common to each building or structure. This subject is covered in detail in Chapter 11 of this text.

Where an "objectionable" flow of current occurs over grounding conductors, Section 250-21 permits taking steps to correct the problem.

Objectionable currents.

The Code does not clarify what is meant by the term "objectionable" for these parallel-path currents nor how much current must flow in the parallel path(s) before it is an "objectionable" current. It seems that an "objectionable" current is one that would prevent a grounding conductor from carrying out its intended function, perhaps from being overloaded. Parallel paths are established where metal raceways are used between buildings or structures and are connected at each end to the metal enclosure. Another common parallel path is established by the Code-required bonding of metal water and gas pipes to the neutral at each building or structure.

One or more steps that can be taken to stop "objectionable" current, provided that "effective grounding" per Section 250-51 is met, include:

1. Discontinue one or more but not all grounding connections.
2. Change the location of grounding connections.
3. Interrupt the continuity of the grounding conductor path.
4. Other remedies acceptable to the Authority Having Jurisdiction.

It is essential to electrical safety that we understand that this section does not permit abandoning all grounding connections to eliminate objectionable currents. Nor does it permit eliminating the equipment grounding conductor connection to equipment. Finally, it does not permit grounding equipment to only a grounding electrode. "Effective grounding," as required by Section 250-51, must be maintained after any and all modifications are made to the grounding connections.

The Code clarifies that temporary currents that flow on the grounding conductors while they are carrying out their intended function, and currents that introduce noise or data errors in electronic equipment such as data processing equipment, are not to be considered "objectionable" currents.

Objectionable Currents

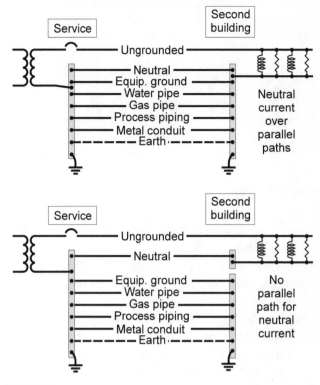

Figure 13-6

Mobile homes and recreational vehicles.

Section 250-44(e) requires that mobile homes and recreational vehicles be grounded in accordance with Articles 550 for mobile and manufactured homes and Article 551 for recreational vehicles.

The basic requirement in Section 550-23(a) is that mobile home service equipment be located remote from the structure but within sight from and not more than 30 feet from the mobile home. There, the grounded (usually a neutral) conductor is grounded to a grounding electrode. Four insulated and color coded conductors, one of which is an equipment grounding conductor, must be run to the mobile home distribution panelboard. See Section 550-24(a).

Section 550-23(a) Exception No. 1 permits the service equipment to be located remote from the mobile home site, if a disconnecting means suitable for use as service equipment is installed within sight from and not more than 30 feet from the mobile home it serves, and grounding at the disconnecting means complies with Section 250-24. This exception permits the two options for grounding to be used as described earlier in this chapter.

Three conductors are permitted to be run from the service to the mobile home disconnecting means where the neutral is grounded, and four insulated

Service Equipment Located Remote From Mobile/Manufactured Home

Remote service

Disconnecting means <= 30 ft.

3- or 4-Wire Feeder

4-Wire Feeder

Manufactured home

Figure 13-7

conductors in compliance with Section 550-24(a) are run to the mobile home. This exception permits the mobile home service equipment to be located at a common location, as is required by some serving electric utilities.

Section 550-23(a) Exception No. 2 permits service equipment to be installed directly on the manufactured home under three conditions.

a. Service equipment is installed in a manner acceptable to the authority having jurisdiction.

b. The service equipment installation complies with Article 230, and

c. Means are provided for the connection of a grounding electrode conductor to the service equipment and routing it outside the structure.

Where this concept is chosen, the service is grounded to the grounding electrode at the manufactured home, and no service disconnecting means remote from the manufactured home is required.

Agricultural buildings.

The grounding of agricultural buildings is covered in Article 547 of the NEC®. Sections 250-24(a), (b) and (c) require that where the building or structure houses livestock, an insulated or covered cop-

per conductor must be installed where the equipment grounding conductor is run underground from one building or structure to another. See Chapter Sixteen of this text for additional information on the subject.

Two buildings in one structure.

Figure 13-8 shows two buildings in one structure that are separated by a fire wall. The two buildings may be served by a single service or by multiple services in accordance with Section 230-2(a). As previously discussed, Section 250-54 requires that all grounding electrodes at a building be bonded together. If two or more grounding electrodes are used, they become, in effect, a single electrode if they are connected together with a conductor properly sized per Section 250-94.

Two Buildings in One Structure

Fire wall for building separation

Bonding grounding electrodes together permitted but not required. 250-54

Figure 13-8

Since the fire wall creates two buildings, bonding of the electrodes together is desirable but not required by Code. The electrodes can be bonded together by connecting them to a common metallic water piping system or by a bonding jumper.

Chapter Thirteen: The questions included here were developed using material included in this chapter. The answers can be found by reviewing the text. It is also important that students make use of the 1996 NEC®, where many answers can be found. See page 237 for answers.

1. A conductor used to connect equipment or the grounded circuit of a wiring system to a grounding electrode or electrodes is defined as a ____.
   a. main bonding jumper
   b. grounding conductor
   c. grounded conductor
   d. neutral conductor

2. Two general methods are provided for grounding electrical systems at separate buildings. One method requires that both the grounded circuit conductor (usually a neutral), and the noncurrent-carrying equipment are to be connected to a ____at the separate building.
   a. isolated ground bus
   b. equipment grounding conductor
   c. grounding electrode system
   d. subpanel

3. The general requirement for grounding at additional buildings is to ground the neutral at the additional building or structure. In this case, the electrical system is treated like a ____ for grounding purposes, although the separate building or structure is supplied from a service by a feeder or circuit.
   a. feeder
   b. isolated power system
   c. subfeeder
   d. service

4. An accepted method for grounding at separate buildings recognizes the installation of an ____ where it is run with the feeder or circuit ungrounded (hot) conductors.
   a. approved conductor
   b. identified bonding jumper
   c. equipment grounding conductor
   d. acceptable conductor

5. Where no grounding electrodes exist at a separate building or structure, one or more grounding electrodes are required to be installed. All of the following are acceptable EXCEPT
   a. ½-inch trade size conduit
   b. ⅝-inch iron or steel not less than 8 feet long
   c. ¾-inch galvanized conduit
   d. plate electrode

6. Where an equipment grounding conductor is run with the feeder or circuit from a first building to a separate building, it is required to be properly sized. The size of this equipment grounding conductor is based upon the ampere rating of the overcurrent protective device ____ the feeder or circuit.
   a. on the load side of
   b. in the transformer of
   c. ahead of
   d. in the service disconnect of

7. A grounding conductor is used to connect the ____ circuit conductor, or equipment grounding conductor and equipment, to the grounding electrode system at the separate building(s).
   a. feeder
   b. main bonding
   c. grounded
   d. branch

8. Under certain circumstances, the grounded circuit conductor ____ to be used for grounding equipment on the line (supply) side of the main disconnecting means for separate buildings.
   a. is permitted
   b. is not permitted
   c. is permitted by special permission
   d. is required

9. Currents that introduce noise or data errors in electronic equipment, such as data processing equipment, are ____ considered objectionable currents.
   a. not to be
   b. considered to be
   c. always to be
   d. sometimes

10. Where the additional building or structure houses livestock, an insulated or covered ____ conductor must be installed where run underground.
   a. bonding
   b. copper clad
   c. aluminum
   d. copper

14

Ground-Fault Circuit-Interrupters

Objectives

After studying this chapter, the reader will be able to understand

- Requirements for replacement of ungrounded receptacles.
- Requirements for locations of GFCI-protected receptacles.
- Ground-fault circuit-interrupter principles of operation.
- Ratings of GFCI devices.
- Markings for GFCI devices.
- GFCI application and consideration.

Shock Hazard Protection. [1]

The Underwriters Laboratories (UL) requirement for Class A GFCIs is that tripping shall occur when the continuous 60-hertz differential current exceeds 6 mA but shall not trip at less than 4 milliamperes (5 mA ± 1 mA). Some people contend that 5 mA is too low and should be increased to 10 mA or higher.

Several eminent investigators, including C.F. Dalziel, F.P. Kouwenhoven, O.R. Langworthy and others, have prepared papers on the dangers of electric shock hazards. They define "let-go" current as the maximum current at which a person is able to release a conductor by commanding those muscles directly stimulated by the shock. Currents over the "let-go" levels are said to "freeze" the victim to the circuit.

Appended to this report is one of Dalziel's many papers, titled "Electric Shock Hazard" and published in the Institute of Electrical and Electronics Engineers (IEEE) Spectrum, Vol. 9, February 1972. It summarizes the studies that estimate shock currents based on the effective impedance of the body under various conditions. According to Dalziel, the reasonably safe electric current for normal healthy adults is the let-go current from which 99.5 percent of the population can extricate themselves from the circuit by releasing the conductor. On page 44, Dalziel states, "[s]o far, it has been impossible to obtain reliable [let-go] values for children; they just cry at the higher values." However, the IIT Research Institute report cites a value for children evidently based on engineering judgment by Dalziel and others. The following summarizes the let-go currents for 99.5 percent of the population:

For children	4.5 milliamperes (mA)
For women	6.0 mA
For men	9.0 mA

Based on information provided in the report by IIT Research Institute, 4.5 milliamperes for children may be the appropriate GFCI trip level relative to the 6 mA and 9 mA thresholds for men and women, but the rationale for this selection is not very evident in Dalziel's or other publications.

Based on Dalziel's data, UL's estimates of the percentage of the population that would be protected against inability to let-go at various current levels is shown in the following table.

For general purposes, a "let-go" limit for GFCIs higher than 6 mA appears inappropriate because too large a fraction of the population would be left unprotected. UL has decided to continue to designate 5 mA (±1 mA) as the limit based on tolerated reaction and physiological effects. Moreover, UL retains the 5 mA limit because the 4.5 mA level is only an estimate that is not based on data and because the 5 mA limit has withstood the test of time with no evidence of being inadequate and has appeared in a number of standards for many years.[2]

Table 14-1
Percentage of the Population Estimated to Be Protected Against Inability to Let Go for Several Levels of Shock Current

Level of Shock Current	6mA(rms)	10mA (rms)	20mA (rms)	30mA (rms)
Men	100%	98.5%	7.5%	0%
Women	99.5%	60%	0%	0%
Children*	92.5%	7.5%	0%	0%

*half of let-go threshold for men

The first sensation of electricity can be felt by most people at currents considerably less than 0.5 mA, 60 Hz (frequency in Hertz). Current near 0.5 mA may produce an involuntary startled reaction such as to cause a person to drop a skillet of hot grease or cause a workman to fall from a ladder. As the current increases, involuntary muscular contractions increase, accompanied by current-generated heat.

Higher currents of longer duration than one second can cause the heart to go into ventricular fibrillation. Ventricular fibrillation is considered the most dangerous effect of electric shock. Once fibrillation begins, it practically never stops spontaneously. Death is almost certain within minutes. The rhythmic contractions of the heart become disordered, its pumping action stops, and the pulse soon ceases altogether. Fibrillation in adults can occur at 52 mA and in children at 23 mA. According to V.G. Biegelmeier, the onset of fibrillation in a 50-kilogram (110 pound) adult occurs within the range of 50 mA to 200 mA when the duration of the shock exceeds two seconds.[3] The table on the following page shows implied safe voltages based on these values.

The table's 3,000-Ω (ohm) body resistance column, for example, indicates that 15 volts would result in a shock current of 52 milliamperes.

Table 14-2
Implied Safe Voltage[4] Based on Several Published Values of Body Resistance and Selected Body Current Safety Criteria as Published by Dalziel

Criterion	Body Resistance 300 ohms	Body Resistance 500 ohms	Body Resistance 1500 ohms	Body Resistance 3000 ohms
"let-go" 4.5 mA for children	1.35 volts	2.25 volts	6.75 volts	13.5 volts
"let-go" 9 mA for adult males	2.7 volts	4.5 volts	13.5 volts	27 volts
fibrillation at 23 mA 5 sec. pulse of 60-Hz current for 18 kg children	6.9 volts	11.5 volts	34.5 volts	69 volts
fibrillation at 52 mA 5 sec. pulse of 60-Hz current for 50 kg adult	15.6 volts	26 volts	78 volts	156 volts

Three-Wire Grounded System vs. GFCI. [5]

For many years, grounding was emphasized as the primary means for the protection of the electrical system, equipment and personnel from fires and injury, as well as for operating and maintenance advantages. The grounded neutral as a protective element was recognized more and more with each succeeding edition of the National Electrical Code® (NEC®). The Code emphasizes "effective" grounding in order to ensure that wiring faults to ground became "overcurrents" as required to activate the overcurrent device. The wiring and equipment were designed to be protected, and the thought that the grounded system also provided adequate protection for people became ingrained and accepted.

The trend toward grounding equipment and appliances has been gradual, characterized as deliberate but cautious. The belief that grounding provided adequate protection against electric shock and fire hazards became so ingrained that consumers generally have not recognized its limitations, and they find it difficult to accept other more effective means of protection from electric shock and electric fire hazards.

Ground faults occur when electrical current flowing in a circuit returns:

- through the equipment grounding wire,
- through conductive material other than the electrical system ground (metal, water, plumbing pipes, etc.),
- through a person, or
- through a combination of these ground return paths.

When a person becomes a pathway for electricity flowing to ground, the person will incur an electrical shock, may be seriously injured or may be electrocuted depending on the current level, current duration, the person's size and the current pathway through the body. On the other hand, arcing ground faults can occur just about anywhere on an electrical system, resulting in a fire. In either case, the ground faults may be of too low a magnitude to activate the overcurrent device and interrupt the flow of electricity.

A person may become a pathway to ground in one of two ways; in series contact or in parallel contact. In series contact, the person is the only current path to ground; the ground wire is not involved. There are many ways in which contact can occur. One example of series contact was that of an infant that stuck a hairpin into a receptacle slot while sitting on a floor heating vent. The infant was electrocuted. Another case involved a man operating a metal-encased electric drill that incorporated a 3-wire grounded cord. He used a "two-to-three" wire adapter but did not connect the adapter pigtail to the wall plate grounding screw. Inadvertently, the pigtail touched the plug blade, thus energizing the drill case and electrocuting the man. In this incident, the equipment ground contributed to the electrocution by providing a current path from the drill case through the body to ground. In both cases, the 3-wire grounded system was totally ineffective since the ground wire was not involved in the current path, and the current through the body was not large enough to trip the overcurrent protective device.

In parallel contact, the victim becomes a path to ground in parallel with the equipment grounding conductor. One scenario of parallel contact occurs when the metal case of an appliance becomes charged electrically by some internal fault condition resulting in current leakage to ground via the grounding conductor. The leakage to ground, however, may not be sufficient to activate the branch-circuit overcurrent protective device. A person who touches the charged case and, at the same time, contacts a grounded surface such as a water faucet or pipe will be subjected to an electric shock. In such parallel contact situations, the effectiveness of the grounding conductor in preventing electrocution of the victim is dependent on several variables, including the following:

- on whether or not the ground-fault current reaches the instantaneous trip level of the overcurrent protective device (which is relatively high—over 15 amperes),
- on how fast the overcurrent device reacts,
- on the voltage level from faulted enclosure to ground, and
- on the impedance of the grounding path (composed of connections, contacts and the ground wire).

An effective grounding system depends upon the integrity of many series connections which must be properly made and maintained. The higher the resistance or impedance of the ground path, the less effectual will be the protection provided by the 3-wire grounded system. Higher ground impedances may be due to long wire lengths, small wire sizes, loose and/or corroded ground wire clamps and connections and other causes. A detailed analysis of the relationship of the impedance ratio of the line circuit to ground circuit to shock current levels is described in the previously cited CPSC paper titled "Three Wire Ground System vs. GFCI" and the cited IITRI report.

A similar study by Mr. R.H. Lee of Dupont Company[6] corroborates IITRI's analysis of circuit impedances. His paper deals with the hazard vs. safeguard of a 3-phase grounded power distribution system.

An assessment of the effectiveness of the 3-wire grounded system for providing protection against electric shock hazards was conducted by Mr. A.W. Smoot of the Underwriters Laboratories.[7] He analyzed 164 fatal electric shock accidents occurring over a 3¾-year period in and around homes. His study indicated the limitations of the 3-wire system and was submitted to NEC® to support proposed GFCI amendments to the 1971 Code.

Principles of operation. [8]

"The GFCI sensing system continuously monitors the current balance in the ungrounded 'hot' conductor and the neutral conductor. If the current in the neutral wire becomes less than the current in the 'hot' wire, a ground fault could exist. A portion of the current returns to the supply by some path other than the neutral wire. With a current imbalance as low as 6 milliamperes, the GFCI will interrupt the circuit and this will be shown by a trip or 'off' indicator on the device." See Figure 14-2.

Ground-Fault Circuit-Interrupter Principles of Operation

*Receptacle-type GFCIs switch both the phase and neutral.

Figure 14-2

"The GFCI does not limit the magnitude of the ground-fault current. It limits the time that a current of given magnitude can flow. The trip level-time combinations are based on physiological data established for avoiding injury to normal healthy persons. These trip level-time combinations may be too high for persons with heart problems, such as those wearing a pacemaker or under treatment in health care facilities."

The principle of operation of ground-fault circuit-interrupters provides a significant advancement in safety for both equipment that is grounded by an equipment grounding conductor as well as for equipment that is ungrounded. Since the GFCI detects an imbalance of current in both the supply and return paths, it protects equipment supplied

GFCI Tripping Characteristics[8]

Figure 14-1

by both a 2-wire circuit and a 2-wire with ground circuit. This is the reason some NEC® sections will allow a grounding-type receptacle to be supplied on a 2-wire circuit that has GFCI protection.

Several kitchen appliances, as well as portable heaters, are manufactured with a 2-wire supply cord and, thus, are ungrounded. A significant advancement in safety is realized where these appliances are supplied from receptacle outlets that have GFCI protection.

GFCI Required for Replacement Receptacles.

The general requirement in Section 210-7 is that receptacles installed on 15- and 20-ampere branch circuits be of the grounding type and have their grounding contacts effectively grounded by an equipment grounding conductor of the supply branch circuit.

Ground-fault circuit-interrupter protection of receptacles is required for receptacles that are replaced in outlets or at locations that are now required by the Code to be GFCI protected. [Section 210-7(d)]

This requires that installers be aware of the rules that call for GFCI protected receptacles in areas covered by Section 210-8 for dwellings, as well as many other locations in the Code for other facilities. This includes dwelling unit bathrooms, garages, outdoor receptacles, etc. Also, this replacement requirement extends to other than dwelling units such as commercial repair garages, elevators and elevator pits, health care facilities and bathrooms in commercial and industrial facilities, as provided in Section 210-8(b).

Table 14-3
GFCI Requirement Locations

Location	NEC® Section
Accessory buildings	210-8(a)(2)
Bathrooms, dwellings	210-8(a)(1)
Bathrooms, nondwellings	210-8(b)(1)
Boathouses	555-3
Construction sites	305-6
Crawl spaces, dwellings	210-8(a)(4)
Elevators, etc.	620-85
Existing	210-7(d)
Fountains	680-51(a)
Garages, dwellings	210-8(a)(2)
Garages, commercial	511-10
Health care facilities	517-20(a)
Hydromassage bathtubs	680-70

Kitchens in dwellings	210-8(a)(6)
Marinas and Boatyards	555-3
Mobile Homes	550-8(b)
	550-23(d)
Outdoors, dwellings	210-8(a)(3)
Rooftops, nondwellings	210-8(b)(2)
RV Parks	551-71
Signs, Outdoor Portable	600-10(c)(2)
Spas and Hot Tubs	680-40
	680-41(a)(2)
	680-41(a)(3)
	680-41(b)(1)
	680-42
Swimming Pools	680-5(b)
	680-5(c)
	680-6(a)(1) Ex.
	680-6(a)(3)
	680-6(b)(2)
	680-20(a)(1)
	680-26(b)
	680-31
Therapeutic Pools and Tubs	680-62(a)
	680-62(f)
Wet Bar Sinks, dwellings	210-8(a)(7)

Both circuit breaker and receptacle-type ground-fault circuit-interrupters are permitted to be used to protect the branch circuit, where nongrounding-type receptacles are replaced with grounding-type

Replacement of Receptacles Where Now Required to be GFCI Protected

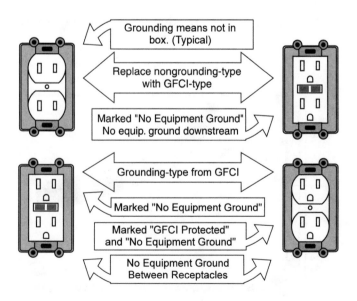

GFCI protection may be circuit breaker or receptacle type.

Figure 14-3

receptacles, where an equipment ground does not exist. These receptacles must be marked "No Equipment Ground" so the user will be informed that even though the receptacle has an equipment ground slot, an equipment grounding conductor is not connected to the device. An equipment grounding conductor is not permitted to be extended from the GFCI-type receptacle to any outlet supplied from the GFCI receptacle.

Where grounding-type receptacles are not on a circuit that supplies an equipment grounding conductor but are protected on the line side by a ground-fault circuit interrupter, the receptacle(s) must be marked "GFCI Protected" and "No Equipment Ground." This will inform the user of the condition of the circuit at the receptacle location. The grounding-type receptacle is protected by the GFCI device, but an equipment grounding conductor is not present as may be required for some equipment such as computers. An equipment grounding conductor is not permitted to be con-

nected between the grounding-type receptacles as this would present incorrect information to the uses. As explained earlier, the GFCI device will protect the user from line-to-ground faults even though an equipment grounding conductor is not connected to the devices.

Receptacles in Bathrooms.

Section 210-8(a) requires that all 125-volt, single-phase, 15- and 20-ampere receptacles installed in dwelling unit bathrooms have ground-fault circuit-interrupter protection for personnel.

The word "bathroom" is defined in Article 100 as "An area including a basin, with one or more of the following: a toilet, a tub, or a shower." It is important to recognize that the definition does not use the word "room" but rather "an area" to describe the location where these receptacles must be provided with GFCI protection. This makes it clear that GFCI protection of these receptacles is required even though the receptacle(s) may be in separate rooms but are in the same "area." No distance measurement is given in the Code beyond which GFCI protection of these receptacles is not required. This definition applies to all bathrooms regardless of type of facility, due to the requirement

GFCI Protection Requirements in Bathrooms

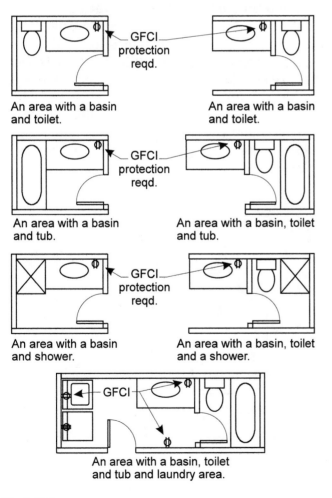

An area with a basin and toilet.

An area with a basin and toilet.

An area with a basin and tub.

An area with a basin, toilet and tub.

An area with a basin and shower.

An area with a basin, toilet and a shower.

An area with a basin, toilet and tub and laundry area.

Figure 14-4

GFCI Requirements for Garages and Grade-level Portions of Unfinished Accessory Buildings

Garage and grade-level portions of unfinished accessory buildings

1. Duplex receptacles require GFCI protection unless they supply two appliances located in dedicated space that are not easily moved from one place to another in normal use and are cord- and plug-connected per Sections 400-7(a)(6), (a)(7) or (a)(8).

2. Single receptacle requires GFCI protection unless it supplies a single dedicated appliance that is not easily moved in normal use and is cord- and plug-connected per Sections 400-7(a)(6), (a)(7) or (a)(8).

3. Receptacle for laundry equipment requires GFCI protection unless it is a single contact device.

4. Receptacle that is not readily accessible does not require GFCI protection. Typical for garage door opener. (Exception No. 1)

5. Additional receptacle required if other receptacles in garage are not readily accessible or are located for appliances that are within dedicated spaces.

Figure 14-5

in Section 210-8(b) for other than dwelling units.

See Figure 14-4 for a graphic illustration of bathroom "areas."

Receptacles in dwelling-unit garages and unfinished accessory buildings.

Section 210-8(a)(2) generally requires that all 125-volt, single-phase, 15- and 20-ampere receptacles installed in dwelling unit garages and in grade-level portions of unfinished accessory buildings used for storage or work areas have GFCI protection.

Exception No. 1 excludes those receptacles that are not readily accessible. "Readily Accessible" is defined in Article 100 as "Capable of being reached quickly for operation, renewal, or inspections, without requiring those to whom ready access is requisite to climb over or remove obstacles or to resort to portable ladders, chairs, etc." No specific height above the garage floor is given as a qualifier. In most cases, receptacles that are located on the ceiling for garage door operators are not considered "readily accessible."

Exception No. 2 to Section 210-8(a)(2) has been revised to exclude from GFCI requirements a single receptacle for a single appliance or a duplex receptacle for two appliances that occupies dedicated space for each appliance, that in normal use is not easily moved from one location to another and that is cord- and plug-connected. This will clarify that duplex or multi-receptacles installed at outlets in a dwelling unit garage must have ground-fault circuit-interrupter protection for personnel, unless a duplex receptacle is used for the supply to two appliances that meets all the conditions.

Dwelling unit outdoor receptacles.

Section 210-8(a)(3) requires all 125-volt, single-phase, 15- and 20-ampere receptacles installed outdoors to have GFCI protection for personnel. This requirement applies to these receptacles that are installed outdoors, for any purpose and at any height above grade or platform and do not have to be accessible from grade level before the requirement applies.

An exception excludes from the GFCI requirement those receptacles that are not readily accessible (defined in Article 100), are on a dedicated branch circuit and are for electric snow-melting or deicing equipment as covered in Article 426. Sec-

GFCI Requirements for Dwelling Unit Outdoor Receptacles

Figure 14-6

tion 426-28 requires ground-fault protection of equipment (not GFCI protection for personnel) for resistance heating elements use for electric snow-melting or deicing.

Dwelling unit crawl space receptacles.

Section 210-8(a)(4) requires that receptacles that are of the 125-volt, single-phase, 15- and 20-ampere type and located in dwelling unit crawl spaces that are at or below grade level have GFCI protection for personnel.

These receptacles are often installed for servicing equipment such as heating, air-conditioning or refrigerating equipment that is located in crawl spaces.

GFCI Requirements for Dwelling Unit Crawl Space Receptacles

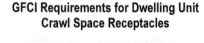

Figure 14-7

Unfinished dwelling unit basements.

Section 210-8(a)(5) requires that receptacles that are of the 125-volt, single-phase, 15- and 20-ampere type and located in dwelling unit unfinished basements have GFCI protection for personnel.

GFCI Requirements for Dwelling Unit Unfinished Basements

1. Duplex receptacles require GFCI protection unless they supply two appliances located in dedicated space that are not easily moved from one place to another in normal use and are cord- and plug-connected per Sections 400-7(a)(6), (a)(7) or (a)(8).

2. Single receptacle requires GFCI protection unless it supplies a single dedicated appliance that is not easily moved in normal use and is cord- and plug-connected per Sections 400-7(a)(6), (a)(7) or (a)(8).

3. Receptacle for laundry equipment requires GFCI

protection unless it is a single contact device.

4. Receptacle that is not readily accessible does not require GFCI protection. Ex. No. 1. (Defined in Article 100)

5. Additional receptacle required if other receptacles in unfinished basement are not readily accessible or are located for appliances that are within dedicated spaces.

Figure 14-8

A definition for "Unfinished Basements" has been included as follows: "Portions or areas of the basement not intended as habitable rooms and limited to storage areas, work areas and the like." It is likely that building codes will be consulted for determining whether rooms in basements are permitted to be classified as habitable rooms.

Exception No. 1 excludes from the GFCI requirement those receptacles that are not readily accessible. "Readily Accessible" is defined in Article 100 as "Capable of being reached quickly for operation, renewal, or inspections, without requiring those to whom ready access is requisite to climb over or remove obstacles or to resort to portable ladders, chairs, etc." No specific height above the garage floor is given as a qualifier.

Exception No. 2 to Section 210-8(a)(5) has been revised to exclude from GFCI requirements a single

receptacle for a single appliance or a duplex receptacle for two appliances that occupy dedicated space for each appliance, that in normal use are not easily moved from one location to another and that are cord- and plug-connected. This will clarify that duplex or multi-receptacles installed at outlets in a dwelling unit garage must have ground-fault circuit-interrupter protection for personnel, unless a duplex receptacle is used for the supply to two appliances that meets all the conditions.

Kitchen counter receptacles.

All 125-volt, single-phase, 15- and 20-ampere receptacles that serve countertop surfaces in dwelling unit kitchens are required by Section 210-8(a)(6) to have ground-fault circuit-interrupter protection for personnel.

GFCI Requirements for Dwelling Unit Kitchens

GFCI protection not required as receptacle does not serve counter-top surface

Receptacle outlets that serve countertop surfaces require GFCI protection for personnel.

Figure 14-9

This requirement does not apply to those receptacles that do not serve countertop surfaces but are installed for other purposes such as for a waste disposal, a trash compactor or a dishwasher. Also excluded from this requirement are wall-mounted receptacles that are installed for equipment such as electric clocks or receptacles that are installed for appliances such as refrigerators where the receptacle is located behind the appliance and does not serve the counter surface.

Wet bar sinks.

Section 210-8(a)(7) requires that all 125-volt,

single-phase, 15- and 20-ampere receptacles installed to serve the countertop surfaces and are located within 6 feet of the outside edge of a wet bar sink in a dwelling unit.

Receptacles that serve wall spaces and are located in the vicinity of the wet bar sink but are not located at or within six feet of the wet bar sink countertop are not required to have GFCI protection.

Rooftop heating, air-conditioning and refrigeration equipment receptacle outlet.

Receptacles installed on rooftops of other than dwelling units must have ground-fault circuit interrupter protection for personnel in accordance with Section 210-8(b)(2).

This receptacle is for the purpose of servicing heating, air conditioning and refrigeration equipment on rooftops, in attics and in crawl spaces.

GFCI Requirements for Construction Sites

GFCI protection required for all 125-volt, 15- or 20-ampere receptacle outlets used for temporary power during construction, remodeling, maintenance, repair, or demolition of buildings, structures, equipment or similar activities.

For receptacles not covered above, the "Assured Equipment Grounding Conductor Program" as provided in Section 305-6(b) is permitted to be used.

Figure 14-11

HACR Rooftop Receptacles

15 or 20 ampere 125 V receptacle
Maximum 25 ft. from HACR equipment
GFCI protection required except dwelling unit(s)
Not connected on load side of disconnect

Figure 14-10

GFCI requirements on construction sites.

Section 305-6 contains requirements for ground-fault protection for personnel using temporary power at buildings, structures or on equipment or similar activities while performing any of the following tasks:
- construction,
- remodeling,
- maintenance,
- repair,
- demolition, or
- similar activities.

Ground-fault circuit-interrupter protection is required for all 125-volt, single-phase, 15- and 20-ampere receptacles used for these purposes. The rule is the same regardless of whether the power is from a temporary construction service, from an existing permanent power source or from one or more receptacles installed as a part of the permanent wiring of the building or structure.

A cord set incorporating listed ground-fault circuit-interrupter protection is permitted to be used by personnel in lieu of permanently or temporarily installed receptacle outlets. This may be the most practical means of complying with this requirement, especially for maintenance and repair activities.

For receptacles like 250-volt, 15- or 20-ampere or 125/250-volt, 30- or 50-ampere, the assured equipment grounding conductor, as provided for in Section 305-6(b), is permitted. This program requires a written procedure enforced by one or more individuals at the construction site. Detailed requirements for testing of the equipment grounding conductor continuity on a specific time schedule are provided. A record of this testing program must be maintained and provided to the authority having jurisdiction.

Application guide for ground-fault circuit-interrupters. [8]
Definitions.

Ambient temperature - The temperature of the surrounding medium that comes in contact with the device or equipment. (For an enclosed device, it is the temperature of the medium outside the enclosure.)

Current interrupting rating - The value of the available RMS symmetrical current at a specified voltage that the device is capable of interrupting under prescribed test conditions.

Electrical shock hazard - A condition, which could result in injury or electrocution to a person, caused by the flow of electrical current through the body.

Endurance test - A test made to determine compliance with specified mechanical and electrical life requirements.

Ground fault - An unintentional electrical path between a part operating normally at some potential to ground, and ground.

Ground-fault circuit-interrupter (GFCI) - A device intended for the protection of personnel. It deenergizes a circuit or portion thereof within an established period of time when a current to ground exceeds some predetermined value that is less than that required to operate the overcurrent protective device of the supply circuit.

1. **Class A GFCI**—A ground-fault circuit-interrupter that is designed to trip when fault current to ground is 6 milliamperes or more.

2. **Class B GFCI**—A ground-fault circuit-interrupter that is designed to trip when fault current to ground is 20 milliamperes or more.

GFCI test - The integral part of a GFCI which checks the ground fault sensing and tripping function of the device.

Ground-fault current - Current that flows through a ground fault.

Ground-fault trip time - the elapsed interval between the time when the ground-fault current is first applied and the time when the circuit is interrupted.

Nuisance trip - Tripping caused by conditions other than those for which the device is intended to respond.

Rated Continuous Current - The maximum rms current which a device or an assembly is designed to carry.

Rated frequency - The frequency for which the device is designed.

Rated maximum ambient temperature - The maximum ambient temperature at which the device is designed to carry continuous current.

Rated voltage - The nominal rms voltage for which the device is designed to be used.

Service conditions - The conditions under which a device is designed to be used.

Trip free device - A device wherein the tripping operation cannot be overridden manually.

Trip threshold - The ground-fault current which, when it is reached, will cause tripping of a GFCI.

Tripping - The automatic opening of a device.

Double Grounded Neutral Protection.

A Class A GFCI is designed so that it will automatically trip if the neutral conductor is grounded on the load side of the sensor.

If a loadside neutral fault to ground and a ground fault occurred simultaneously, some of the fault current would flow through the neutral wire to the sensor, and some would flow through the inadvertent ground path. If such a ground connection occurred, it would be possible for a person to contact a hot wire and ground, having the ground-fault current flow through the inadvertent neutral ground and the neutral to the service entrance. Under this condition, there may not be enough imbalance in current through the sensor to cause the GFCI to trip.

The double grounded neutral protection feature prevents this condition from occurring.

This protective feature demands proper loadside building wiring. It continuously monitors the loadside wiring to detect a double-grounded neutral condition.

Test means.

Each GFCI is equipped with a means of testing the ground-fault trip capability. A simulated ground fault through a test resistor is applied to the sensor through the activation of the test means. It is important that user testing procedures be followed.

GFCI tripping characteristics.

Typical trip characteristics of GFCIs are shown in Figure 14-1.

Types of GFCIs.
Basic GFCI.

The basic GFCI consists of a ground-fault detecting means coupled with a circuit-interrupting means.

The detecting means measures the ground-fault current as the difference between outgoing and incoming load current on the protected conductors. The current difference is suitably processed to activate the circuit-interrupting means.

The circuitry module and associated interrupting means which make up the basic GFCI should be mounted in an enclosure or combined with some other device in order to be suitable for installation on a wiring system. The basic GFCI is combined into a complete device to make up the various types described below.

Blank-face type GFCI.

The blank face type GFCI has no receptacle openings. It provides both personnel ground-fault protection and switching capabilities. It may be rated for motor switching if so marked. It is not intended to be used as a substitute for a wall-mounted switch.

This type of device may have flexible insulated leads or binding screw terminals or push-in terminals and is to be mounted in an enclosure such as an outlet box. Means for five field connections are provided: line, line neutral, load, load neutral and equipment grounding.

This device may be mounted in listed industrial control panel (UL 508, Industrial Control Equipment) and may be used as a motor controller if so marked. It may be mounted in other equipment and be listed by UL in combination with the equipment.

Permanently-mounted type of GFCI.

The permanently-mounted type of GFCI has an integral enclosure as a part of the total unit and is designed to be permanently wired into the supply circuit.

Circuit breaker type.

The circuit breaker-type GFCI incorporates within one device the ability to provide all the functions of a circuit breaker, as well as ground-fault protection for personnel for the entire circuit.

It is intended to be mounted in an enclosure and frequently may be installed as a direct replacement for a circuit breaker in a load center or a panelboard.

Circuit breaker-type GFCIs are available as single-pole and two-pole units with various plug-in and lug type electrical connections. Line neutral connections are usually made by means of a wire lead, and care should be taken to install the device according to the manufacturer's instructions.

Receptacle-type GFCIs.

The receptacle-type GFCI incorporates within one device one or more receptacle outlets, protected by the GFCI. It is available as a "nonfeed-through" unit which provides GFCI protection only to the outlets in the unit, and as a "feed-through" unit which provides GFCI protection to loadside receptacles as well. The feed-through unit may also be used as a nonfeed-through unit.

Receptacle-type GFCIs are available with various terminating means, including flexible insulated leads and binding screw terminals or push-in terminals. The feed-through unit provides for five field connections: line, line neutral, load, load neutral and equipment grounding. The nonfeed-through unit provides connections for line, line neutral and equipment grounding.

Receptacle-type GFCIs are intended to be mounted in an enclosure such as an outlet box.

Portable type GFCI.

The portable-type GFCI is an easily transportable, self-contained, enclosed GFCI with one or more integral receptacle outlets protected by the GFCI module. In lieu of an integral receptacle outlet, the load conductors may be attached by a cord and connector. Portable type GFCIs are normally connected to a supply receptacle by an integral attachment plug or a plug and cord. Receptacle cord connectors and plugs are available in various general purpose and locking configurations.

The enclosures for these units may be of the indoor type or the outdoor type. Indoor types usually have nonmetallic enclosures. The outdoor types may have metallic or nonmetallic enclosures suitable for the environment in which they are used. (See NEMA Standards Publication No. 250, *Enclosures for Electrical Equipment (1000 Volts Maximum),* for enclosure information.)

The portable-type GFCI has a no-voltage release feature which will automatically interrupt the circuit when any of the supply conductors are open.

Cord-connected type.

A cord-connected type GFCI consists of an attachment plug containing a GFCI, and a connected cord. It provides GFCI protection on the loadside for the cord and equipment to which the cord is attached.

A cord-connected type may be part of other equipment and is listed by UL in conjunction with the equipment.

The cord-connected type GFCI also has a no-voltage release feature which will automatically interrupt the circuit when any of the supply conductors are open.

Rating information.

The rating of a GFCI includes the following: class, rated voltage, rated frequency, rated continuous current and, where applicable, current-interrupter rating.

Class.

The GFCI designated "Class A" will trip when there is a fault current to ground of 6 milliamperes or more. The GFCI designated "Class B" will trip when there is a fault current to ground of 20 milliamperes or more "Class B" GFCIs should be used only for underwater swimming pool lighting installed prior to May 1965 (prior to the adoption of the 1965 National Electrical Code®).

Rated voltage.

The common alternating-current voltage ratings of GFCIs are 120, 120/240, and 240 volts, single phase; and 208Y/120 and 240 volts, 3-phase.

Rated frequency.

The rated frequency of GFCIs is 60 Hz.

Rated continuous current.

The common continuous current ratings of GFCIs are 15, 20, 25, 30, 40, 50 and 60 amperes.

Rated withstand current.

Receptacle, blank-face, portable and cord-connected types have current-withstand ratings. Typically, they are 2,000 amperes, rms symmetrical. Permanently-mounted types normally have current-withstand ratings of 5,000 amperes, rms symmetrical.

Rated interrupting current.

Circuit breaker-types normally have current-interrupting ratings of 5,000, 10,000, or 22,000 amperes, rms symmetrical.

Markings.
General.

GFCIs are marked with the manufacturer's name or trademark, voltage rating and load capacity in amperes. Additional information may also be marked on the GFCI or included with the manufacturer's literature.

Types which are permanently connected into circuit.

For types which are permanently connected into a circuit, which include the switch-type, permanently-mounted-type, circuit breaker-type, and receptacle-type, the following additional information should be provided:
1. Installation instructions,
2. Test instructions to operate upon installation and at least monthly,
3. Test reminder with provisions for checking off monthly tests,
4. Warning against use in wet locations,

5. Warning that GFCI does not protect against shock hazard when both conductors are touched,
6. Statement— "For Use on Grounded Circuits Only,"
7. Class (A) or (B), and
8. Proper conductor termination instructions.

Marking on the device or literature packaged with the device should include the following or equivalent statement: "To avoid accidental bypassing of protection, the load-circuit wiring should be separated from other wiring by suitable insulation, barriers, or restraints."

Class A devices.

Marking on a Class A GFCI or literature packaged with the device should include the following or equivalent statement:

"To minimize false tripping do not connect to swimming pool equipment installed before adoption of the 1965 National Electrical Code®."

Class B devices.

Marking on a Class B GFCI should include the following or equivalent statements:

These instructions may appear in the literature packed with the GFCI.

"Use only with underwater swimming pool lighting fixtures:

CAUTION—To avoid possible shock, disconnect power before servicing fixture. To minimize false tripping, do not connect to more than __ feet of load conductor for the total one-way run."

Portable and cord-connected types.

For portable and cord-connected types which are cord connected, the following additional information should be provided:
1. Load capacity in watts.
2. The word WARNING and a statement to the effect that, to ensure protection against electric shock the test instructions should be followed before an appliance is plugged into any receptacle on the device. The marking describes the significance of the test and informs the user that, in the event of an indication of improper functioning, the cause of the malfunction should be corrected before further use of the device.
3. Unless a cord-connected GFCI has been found suitable for use in wet locations, it should be marked "Do not use where water is likely to enter case" or the equivalent.

4. A portable or cord-connected device should be marked with a statement to the effect that the device does not guard against electric shock resulting from (1) some possible defects or faults in an extension cord or other wiring supplying the GFCI, or (2) contact with both circuit conductors.

5. Unless otherwise marked, a cord-connected GFCI not provided with a permanently attached cord 6 feet or longer in length should be marked "This product should be used only with a three-phase conductor, 120 volts, ___*** ampere supply cord set employing Type ST, SO, STO, SJT, SJO, or SJTO cord. In the event of cord set damage, it should be replaced only with an equivalent cord set" or similar wording.

***15, 20, or 30 ampere, to be specified by manufacturer.

GFCI application.

In most cases tripping of a GFCI is an indication of a dangerous condition that needs correction. GFCI tripping under such conditions is not "nuisance" tripping. The cause of the tripping should be identified and corrected.

Unless the device is a circuit breaker type GFCI, it will not protect the circuit conductors against overcurrent. Separate overcurrent protection must be provided.

Grounding and GFCI protection are used to complement one another, not to replace one another.

Periodic tripping tests should be made in accordance with the manufacturer's recommendations, using the test means on the unit. When the test button is pushed, a predetermined value of ground-fault current is supplied to the GFCI, tripping the unit and thereby testing the GFCI.

To avoid damage to the GFCI, it must be disconnected before feeders or branch circuits are subjected to a megger, high voltage, or hi-pot test. Disconnect the electrical power supply and load conductors to isolate the GFCI.

Carefully read the manufacturer's instructions before installation of the GFCI.

A GFCI does not protect a person who comes in contact with two "hot" wires or any "hot" wire and the neutral wire.

A GFCI does not protect a person from feeling and reacting to a shock.

When using a feed-through receptacle-type GFCI, be certain that outlets to be protected are all on the load terminals or wires of the GFCI.

When a feed-through receptacle-type GFCI is used only to protect its own receptacle, the unused load wire connections, if any, should be properly insulated to prevent electrical hazard.

When a feed-through receptacle-type GFCI is used to protect a complete load circuit or only a portion of a load circuit, the "load" wires should be connected to the remaining receptacles in the branch. THE LINE AND LOAD CONNECTIONS MUST NOT BE REVERSED; OTHERWISE THE GFCI'S OWN RECEPTACLE WILL BE UNPROTECTED AND WILL REMAIN ENERGIZED AFTER THE UNIT HAS TRIPPED.

Depending upon where it is installed on a branch circuit, a feed-through receptacle-type GFCI can protect all of the receptacles on the circuit, a portion of the receptacles on the circuit, or only its integral receptacle.

Suitable enclosures should be used for GFCIs or receptacles protected by GFCIs. This is especially important in the presence of moisture or a corrosive atmosphere.

The GFCI contains an electronic circuit which is designed to be resistant to electrical interference. In the same way that a television set or radio suffers from momentary "static" in the presence of interference, so may the GFCI. However, in the television or radio, the interference is merely a temporary nuisance. The GFCI, as a safety device, may interpret this interference as a ground fault and trip. Therefore, the GFCI should not be used in an application where continuity of service takes precedence over the personnel protection feature, such as life-sustaining equipment in health care facilities.

The trip characteristics established for the GFCI ground fault sensor are based on physiological data taken for normal healthy persons. These levels may be too high for persons with heart problems such as those wearing pacemakers or those under treatment in health care facilities.

Portable or cord-connected GFCI types will not protect the receptacle or any portion of the branch circuit into which it is plugged. It will protect only those conductors and devices connected to its load side.

Equipment with high temperature-sheathed heating elements, such as dishwashers and electrical space heaters, tend to have high electrical leakage which may cause tripping when used with GFCIs. This type of equipment should be effectively grounded.

Permanently-connected lights, without receptacles, need not be protected by GFCIs.

Stationary appliances, such as sump pumps and clothes washers, should be effectively grounded. These types of appliances may have high electrical leakage caused by moisture. Refrigerators and freezers, particularly older units, may also have this type of leakage. When these appliances are on a GFCI-protected circuit, the leakage currents could exceed the GFCI trip threshold and cause the GFCI to trip.

High moisture conditions may result in electrical leakage of portable tools. This will cause the GFCI to trip in some instances even if the tool is turned off, if the leakage exceeds the GFCI trip value.

High moisture conditions may result in leakage of unprotected extension cord connectors and aged cord insulation. Replace or repair any aged wiring, extension cord, or portable tool which, under high moisture, causes the GFCI to trip.

GFCI protection cannot be used for ranges, ovens, cooking units, and clothes dryers when the frame of the appliance is connected to the grounded circuit conductor.

If a GFCI is used with a small portable generator, the GFCI circuit should be turned off when starting and until the generator has attained constant speed and stabilized output voltage.

A GFCI feeding an isolation transformer, which supplies circuits such as underwater swimming pool fixtures, will not detect a ground fault on the secondary side of the transformer.

GFCIs should be used only on systems with a grounded circuit conductor. A GFCI can be used on a 2-wire circuit without a ground (green) wire as long as the circuit has a grounded circuit conductor. It should be noted that the equipment ground conductor "green wire" is completely independent of the GFCI circuitry.

The National Electrical Code® requires that the white (or grounded) conductor be grounded only at the service entrance. If this conductor is grounded again on the load side of the GFCI, tripping will occur.

A 120/240V two-pole circuit breaker-type GFCI need not have the load side neutral connected for 240V rated loads.

Certain portable voltage testers or test lamps, when connected between line and ground on the load side of the GFCI, may cause the GFCI to trip.

Some office and laboratory equipment (e.g., personal computers) exhibit ground leakage currents due to designed-in features such as RFI filtering and transient voltage surge suppression. While this current is permitted by applicable product standards, it can result in tripping of a GFCI. This possibility is increased when two or more pieces of equipment, each with some leakage current, are connected to the same branch circuits.

After a GFCI has tripped and the cause has been determined to be a result of this accumulated permitted leakage current, the number of such devices on one branch circuit must be reduced. It may be necessary to provide dedicated circuits or individual GFCI receptacles if this inherent leakage is high.

The equipment manufacturer should be contacted if there is any question pertaining to the expected leakage current level.

GFCIs are specifically tested for immunity to radio frequency noise. See NEMA PP 1, *Procedure for Evaluating Ground-Fault Circuit-Interrupters for Response to Conducted Radio Frequency Energy.*

Circuit breaker or receptacle-type GFCIs are not intended to be field wired to an attachment plug to create a portable GFCI. Such an arrangement will not provide the full protection afforded by a portable or cord-connected GFCI.

References:

[1] *A. Albert Biss,* Ground-Fault Circuit-Interrupter (GFCI) Technical Report, *(Washington, D.C.: U.S. Consumer Product Safety Commission, February 28, 1992).*

[2] *Walter Skuggevig,* 5-Milliampere Trip Level for GFCIs, *(Northbrook, IL: Underwriters Laboratories, March 1989).*

[3] Electric Shock Prevention, *p. 25.*

[4] *From the content of the IITRI report, the voltage and current values are assumed to be "root-mean-square" (RMS) values.*

[5] *A. Albert Biss,* Three-Wire Grounding System vs. GFCI, *(Washington, D.C.: U.S. Consumer Product Safety Commission, December 1985).*

[6] *Ralph H. Lee, Electrical Grounding: Safe or Hazardous?,"* Chemical Engineering, *July 28, 1969.*

[7] *A.W. Smoot,* Analysis of Accidents, *(Northbrook, IL: Underwriters Laboratories, 1971).*

[8] *Application Guide for Ground-Fault Circuit-Interrupters, Standards Publication No. 280-1990, National Electrical Manufacturers Association, 1300 North 17th Street, Suite 1847, Rosslyn, VA 22209. Reprinted with permission.*

Chapter Fourteen: The questions included here were developed using material included in this chapter. It is also important that students make use of the 1996 NEC®, where many answers can be found. See page 237 for answers.

1. Where existing 125 volt, single-phase, 15- and 20-ampere rated receptacles are replaced, or where new receptacles are installed in locations that are required by the Code to be GFCI protected, they must be ___.
   a. of the grounding type
   b. GFCI protected
   c. on dedicated circuit
   d. polarized

2. Circuit breaker or receptacle-type ground-fault circuit-interrupters ____ be used to protect the receptacle, where nongrounding types of receptacles are replaced with grounding type receptacles where an equipment ground does not exist in the receptacle enclosure.
   a. are permitted to
   b. cannot
   c. must always
   d. by special permission can

3. Where used for a single cord- and plug-connected appliance that occupies a dedicated space in a residential garage, GFCI protection is not required where a ____ receptacle is installed and dedicated for use by that appliance.
   a. duplex
   b. single
   c. isolated
   d. 15-ampere rated

4. Portions or areas of a dwelling unit basement that are not intended to be used as habitable rooms and limited to storage areas, work areas and the like, defines:
   a. a basement
   b. a finished basement
   c. an unfinished basement
   d. a storage space

5. Where installed in dwelling unit kitchens, all 125-volt, single-phase, 15- and 20-ampere receptacles that are installed ____ must be ground-fault circuit-interrupter protected.
   a. within 6 feet of the sink
   b. on the dishwasher circuit
   c. to serve counter surfaces
   d. to serve refrigeration equipment

6. The distance of the required receptacle on a rooftop, other than at one- and two-family dwellings, must be located to be not further than ____ feet from the heating, air-conditioning and refrigeration equipment.

   a. 10
   b. 15
   c. 25
   d. 50

7. Where installed in ____ of commercial, industrial and all other nondwelling-type occupancies, all 125-volt, single-phase, 15- and 20-ampere receptacles must be ground-fault circuit-interrupter protected.

   a. kitchens
   b. basements
   c. closets
   d. bathrooms

8. Where installed on roofs of other than ____, all 125-volt, single-phase, 15- and 20-ampere receptacles must be ground-fault circuit-interrupter protected.

   a. one family dwelling units
   b. two family dwelling units
   c. dwelling unit garages
   d. all the above

9. With a current imbalance as low as ____ milliamperes, a GFCI will interrupt the circuit and this will be shown by a trip or "off" indicator on the device.

   a. 3
   b. 4
   c. 5
   d. 6

10. A GFCI of the Class ___ type is designed so that it will automatically trip if the neutral conductor is grounded on the load side of the sensor.

   a. A
   b. B
   c. C
   d. D

11. The basic GFCI consists of a ground-fault detecting means that is coupled with a circuit-interrupting means. This detecting means measures the ground-fault current as the difference between outgoing and incoming load current on the ____ conductors.

   a. protected
   b. neutral
   c. bonded
   d. service

12. A condition which could result in injury or electrocution to a person, caused by the flow of electrical current through the body, is defined as:

   a. a high resistance ground
   b. an electrical shock hazard
   c. a short circuit
   d. ground fault

13. To determine compliance with specified mechanical and electrical life requirements ____ is made.

   a. an endurance test
   b. a ground fault test
   c. an operational test
   d. an overcurrent test

Chapter 15

Equipment Ground-Fault Protection Systems

Objectives

After studying this chapter, the reader will be able to understand

- Requirements for ground-fault protection of equipment and various systems.
- Requirements for feeder ground-fault protection.
- Ground faults in an ungrounded system.
- Testing of equipment ground-fault protection.

Definition.

Ground-fault protection of equipment: "A system intended to provide protection of equipment from damaging line-to-ground fault currents by operating to cause a disconnecting means to open all ungrounded conductors of the faulted circuit. This protection is provided at current levels less than those required to protect conductors from damage through the operation of a supply circuit overcurrent device."* (Article 100)

Protection required.

Section 230-95 of the NEC® requires ground-fault protection of all solidly-grounded wye electrical services of more than 150 volts to ground, but not exceeding 600 volts phase-to-phase for each service disconnect rated 1,000 amperes or more. These provisions do not apply to fire pumps or continuous industrial processes where a nonorderly shutdown would introduce additional or increased hazards.

Ground-Fault Protection Required

Solidly grounded.
> 150V to Gnd.
<= 600V Ø to Ø.
>= 1000A service disconnect rating

Figure 15-1

The maximum setting of the devices is 1,200 amperes, and the maximum time delay is one second for ground-fault currents equal to or greater than 3,000 amperes.

The magnitude of current which will flow in the event of a line-to-ground fault on a grounded system is usually determined by the reactance of the grounded apparatus, the reactance of the lines or cables leading to the fault and the resistance and reactance of the ground return path including any intentional grounding resistance or

reactance. For interconnected systems, calculation of the current may be rather complicated. For simpler cases, an approximation of the available fault current may be obtained.

This equipment protection is required due to a history of destructive burndowns of electrical equipment operating at this voltage level. An electric arc, which generates a tremendous amount of heat, is readily maintained at 277 volts to ground. Since it is not a solid ground connection, and the current is limited by the resistance of the arc, often, not enough current will flow in the circuit to allow the overcurrent device ahead of the fault to open. A great deal of damage is done to the electrical equipment while the arc is burning.

There are basically two types of equipment ground-fault protection systems in use although these systems may have different names in the industry. The most common types are zero-sequence which may have more than one form and residual type which is sometimes referred to as ground-strap or ground-return type. Both types are designed to protect downstream equipment from destructive arcing burndowns. Note that this equipment will not protect equipment or the system on its line side from line-to-ground faults as this fault current will not pass through the ground-fault equipment.

Residual-type system.

The residual or ground-strap type of equipment ground-fault protection system consists of a current transformer, control relay and, usually, a shunt trip circuit breaker. The main bonding jumper

Residual EGFP System

Neutral bar
Current trans
Main bonding jumper
Equipment ground bar
Relay
Fault in equipment
Service enclosure

Figure 15-2

passes through the current transformer as shown in Figure 15-2.

The advantage of this system is it is probably the least expensive, but it is limited to application at the main service. It cannot be used downstream of the main bonding jumper.

With the connections as shown, the maximum ground-fault current will flow through the main bonding jumper and will, thus, be recognized by the current transformer. Although there are many parallel bypass circuits for ground-fault current to flow, the relative impedance of the main circuit is so low compared to all the parallel bypass circuits that a minimum of 90 percent of the total ground-fault current will flow in the main bonding jumper and, thus, be seen by the current transformer.

If a current transformer of properly-selected value is placed in the main bonding jumper and has its secondary feeding a magnetic current relay, we may adjust the relay to close a contact at any desired value of fault. The current transformer value is selected on the basis of the smallest current transformer whose short-time rating will not exceed the maximum value of ground-fault current that may be anticipated.

Examination of the diagram Figure 15-2, describing the ground-fault protective device, will show that the main ground-fault current path is from the transformer to the service, through the feeder to the fault, back to the service over the equipment grounding conductor through the main bonding jumper where it returns to the transformer by the neutral. The parallel circuit from the grounding electrode at the service to the grounding electrode is both a high-resistance and a high-reactance circuit. As a result, little current will return through the earth.

Some small ground-fault current will be carried by the building steel. It is preferable to adequately ground the building steel using the same grounding electrode as used for the service. This prevents the building steel from rising to a dangerous potential above ground and often serves as a grounding electrode as required by Section 250-81(b). Even though the building steel represents a parallel path for fault current to flow, most of the current will flow back to the transformer through the neutral because of the lower reactance of that path as compared to the reactance of the other parallel paths.

Zero-sequence transformer-type system.

Probably the most popular and common type of equipment ground-fault protection system is the zero-sequence type. This system is shown in Figure 15-3 and Figure 15-4. It consists of a current transformer that is placed around all the conductors of the circuit, including the grounded (neutral) conductor, a control relay and usually a shunt trip circuit breaker. As shown, the current transformer must be placed downstream from the main bonding jumper. The equipment grounding conductor does not pass through the window.

Zero Sequence EGFP System

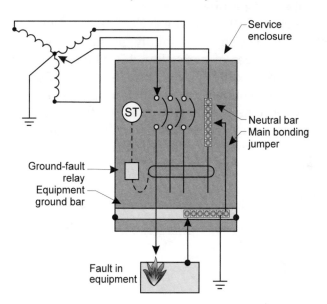

Figure 15-3

Zero Sequence EGFP System

Figure 15-4

The general difference between the types of zero sequence equipment shown in Figures 15-3 and 15-4 is the current transformers are generally built into the circuit breaker shown in Figure 15-4 with an external current transformer through which the neutral passes that may be field installed. Often, these current transformers are used by the circuit breaker as a part of its internal operational system. The equipment ground-fault protection system shown in figure 15-4 is sometimes referred to as a residual type as the ground-fault relay or system sums up the current through all four coils and considers and excess current as "residual." Generally, the current transformer through which all conductors of the circuit pass as shown in Figure 15-3 is installed by the manufacturer of the switchboard. Cables for feeders are field installed.

Under normal operation, the vector summation of all phase and neutral (if used) currents approaches zero. This is due to the cancelling effect of the currents in the conductors. A sensor (differential current transformer) around the phase and neutral conductors detects the current imbalance when a ground fault occurs. This is due to the fault current passing outside the current transformer window(s) which sets up the imbalance which is detected by the system.

The output of the sensor is proportional to the magnitude of the ground-fault current. This output is fed to a ground-fault relay. The relays are usually field adjustable. Pickup ranges of from 4 to 1,200 amperes are common. Time delay settings are available from instantaneous (1.5 cycles) to 36 cycles delay. When the ground-fault current exceeds a pre-selected level, the relay will activate the circuit-interrupting device, which usually is a shunt trip circuit breaker, to open the circuit.

Ground-fault protection not required.

Ground-fault protection of equipment is not required for the service for a continuous industrial process where a nonorderly shutdown will introduce additional or increased hazards. (Section 230-95 Exception No. 1)

Likewise, ground-fault protection of equipment is not required for fire pump services. (Section 230-95 Exception No. 2) Section 700-26 excludes the alternate source of power from the requirement to have ground-fault protection of equipment with automatic disconnection means. This applies to fire pumps that are classified as an emergency system in accordance with Section 700-

1. Indication of a ground fault of the emergency source is required to comply with Section 700-7(d). The indication of the ground fault in a solidly grounded wye emergency system of more than 150 volts to ground and circuit protective devices rated 1000 amperes or more. The sensor is required to be located at, or ahead of, the main system disconnecting means for the emergency source. Instructions on the course of action to be taken must be located at or near the sensor location.

A similar exclusion from the ground-fault protection of equipment requirement is provided for legally required standby systems in Section 701-17. No such exclusion is provided for optional standby systems installed in accordance with Article 702.

System coordination.

System coordination with circuit breaker systems is easily accomplished by use of ground-fault protective devices. These ground-fault protective devices may be cascaded where the economics of the design warrant doing so. The time-delay settings may become lower and lower as we go farther and farther from the service, so that the last ground-fault protective device in the system may even be set for instantaneous trip. Similar coordination also can be obtained by locking out the relay or relays upstream from the sensor that sees the fault first.

System Coordination With EGFP Equipment

Figure 15-5

Since we do not know where a fault-to-ground or enclosure may originate in the system, and since we want to coordinate the protection and clear the fault nearest to its point of origin, it will be necessary to delay the action of tripping the main overcurrent device. The magnetic current relay is thus used to energize a time delay relay to provide proper coordination and maximum continuity of service consistent with safety.

When a fault-to-ground or enclosure occurs, we should not completely discard the possibility, although it may be rare, that the fault may reach a high value. If so, we would want to give the normal main overcurrent device time to function and clear the fault through that means. Also, since we want to set the ground-fault protective relay at as low a value as possible, we want to delay action to permit an overcurrent device near the point of origin of the fault to clear first. This ground-fault protective device is an adjunct to, but does not replace, the main overcurrent device. It gives protection below the rating of the main overcurrent device. Being affected only by ground-fault currents, it does not interfere with normal operation of the main overcurrent device. In effect then, the ground-fault protective device will keep all ground faults to a limited time, selected on the basis of experience, to give maximum safety without undue interruption of service.

Second level required health care facilities.

Special rules for ground-fault protection of electrical systems apply to health care facilities. Section 517-17 requires that where ground-fault protection is provided on the service or feeder disconnecting means, a second level of ground-fault protection be provided in the next level of feeder downstream. This rule clarifies that regardless of whether or not the health care facility is supplied by a service or feeder, where equipment ground-fault protection is provided at the service or feeder disconnecting means, a second level is required downstream.

The additional level(s) of ground-fault protection is not permitted to be installed:

(1) on the load side of an essential system transfer switch, or

(2) between the on-site generating unit(s) described in Section 517-35(b) and the essential system transfer switch(es), or

(3) on electrical systems that are not solidly grounded wye systems with greater than 150 volts to ground, but not exceeding 600 volts phase-to-phase.

Health Care Facilities EGFP Systems

Figure 15-7

This feeder protection is required to be 100 percent selective so that where a ground fault occurs downstream from the feeder overcurrent device, only the feeder overcurrent device will open and the service or feeder main will remain closed. This, of course, is to prevent the blackout of a facility caused when a main would open for a fault that can be isolated to a single feeder. To achieve this coordination, a six cycle, or greater, separation between the service and feeder tripping times is required. Each level of the ground-fault protection system must be tested when first installed.

Feeder ground-fault protection.

Figure 15-8 illustrates a case where equipment ground-fault protection is required for feeders. Section 215-10 requires this protection for feeders that are of the same voltage and current rating as similar services.

This covers the situation where the electrical system is delivered at more than 600 volts and has the service disconnect at that level. A transformer then reduces the voltage of the feeder to that where equipment ground-fault protection

Second Level of EGFP in Health Care Facilities

Additional level of equipment ground-fault protection not permitted:

(1) on the load side of an essential electrical system transfer switch, or

(2) between the on-site generating unit(s) and the essential electrical system transfer switch(es), or

(3) on systems not solidly grounded wye systems more than 150 volts to ground and less than 600 volts phase-to-phase.

Life safety branch
Critical branch
Equipment system

Hospital essential electrical system

Figure 15-6

Equipment Ground Fault Protection for Feeders

Over 600 volt service
Secondary is:
Solidly grounded
More than 150 volts to ground
Less than 600 volts phase-to-phase

Figure 15-8

Equipment Ground Fault Protection for Feeders

Solidly grounded wye, > 150V to ground
<= 600V Ø to Ø, Rated >= 1000 A

Equipment ground-fault protection required at building or structure disconnecting means except for:
1. continuous industrial process ...
2. fire pumps
3. where disconnecting means protected by service or feeder ground fault protection.

Figure 15-9

is required. The risk of destructive arcing burndowns is the same regardless of whether the system is a service or feeder.

An exception excludes feeders that have ground-fault protection on the line side. However, care must be exercised to be certain that each new system created by a separately derived system has the protection required.

Ground-fault protection of building or structure main disconnecting means.

Section 240-13 requires ground-fault protection of equipment to be installed in accordance with Section 230-95 for systems of identical voltage and amperage rating to that for services. Three exceptions are provided. Exception No. 1 excludes a continuous industrial process where a nonorderly shutdown will introduce additional or increased risks. Exception No. 2 excludes fire pumps from the requirements and exception No. 3 excludes the disconnecting means installation from the ground-fault protection requirements where the service or feeder provides the necessary protection.

It is most important that an overall engineered system approach is taken where more than one building or structure is on the premises that are fed from a common service. This is especially important where transfer switches and alternate power sources are installed in the premises wiring system. In most cases, four-pole transfer switches are installed so as to allow the system to function properly without grounding the neutral downstream from the service. It is critical to ground-fault protec-

tion systems functionality that the neutral be completely isolated from any grounding connections downstream from the service.

Combination systems.

An improvement in the protection of distribution systems of 600 volts or less may be obtained by using a ground-fault protective device as a supplement to high-interrupting capacity current-limiting overcurrent protection. Fault currents in low ranges may be recognized and virtually all fault currents can be held to a limited time duration with the resultant increase to the safety of such systems.

If circuit breakers are used as the disconnecting means and overcurrent protection on such secondary distribution systems and a ground-fault reaches a high enough value, the circuit breaker will open and clear all three phases, thus preventing a feedback into the ground fault from the two unaffected phases. When a circuit breaker is used, a ground-fault protection device also may be applied to automatically open the circuit breaker in a reasonable and limited time if a ground fault should develop and not be cleared by some overcurrent device near the point of fault.

If a manually-operated, fused load-break switch is used as the disconnecting means and overcurrent protection, and if a ground fault reaches a high enough value to clear one fuse, an alarm should be provided to indicate a blown fuse so all three phases may be cleared manually.

If an electrically-operated, fused load-break switch is used, it can be provided with an automatic blown-fuse indicator which will immediately open the switch and, thus, all three phases, if any one or more fuses should blow. An electrically-operated switch also permits the application of a ground-fault protective device which will automatically open the switch in a reasonable and limited time if a ground fault should develop and not be cleared by some overcurrent device near the point of fault.

For systems with large available short-circuit current, and which require the best degree of protection, the electrically-operated switch or circuit breaker can provide full protection where it is equipped with high-interrupting capacity current-limiting fuses, a ground-fault protective device and an automatic blown-fuse indicator.

Testing of system.

Section 230-95(c) requires that ground-fault protection systems be performance-tested when first installed on site to ensure that they will operate properly. Experience has shown that the majority of these systems do not operate properly, or at all, when first installed. This is often due to improper field wiring of the system.

The test must be performed in full compliance with the manufacturer's written instructions. These instructions must be furnished with the equipment. A vital part of the test is to remove the neutral disconnect link and megger the neutral to be certain that it is clear from any ground downstream from the service. Accidental or intentional grounding of the neutral past the service will desensitize the ground-fault protection system and render it ineffective.

A written record of these tests must be made and provided to the inspection agency. Safety demands that the equipment be properly installed, as does the threat of lawsuit where negligence of the installer can be proven.

Optional protection.

Figure 15-10 illustrates an installation where equipment ground-fault protection is not required even though the system voltage is the same as that where it is required. The difference is the rating of the disconnecting means is not 1,000 amperes or greater, even though the combined rating exceeds 1,000 amperes. For example, a service with a single 2000 ampere main requires equipment ground-fault protection at the stated voltage and ampere rating. If a service with four, six-hundred ampere overcurrent devices is installed rather than the

Optional Equipment Ground Fault Protection

Solidly grounded wye, > 150V to ground
<= 600V Ø to Ø

Figure 15-10

single main, equipment ground-fault protection is not required. Obviously, the system can benefit from equipment ground-fault protection even though not required by the Code.

In the case where a 900-ampere fuse is installed in a 1,200-ampere switch, the requirements are different. Equipment ground-fault protection is required since the fuse holder will accept a 1,000-ampere fuse. Where adjustable circuit breakers are used, the rating, for purposes of Section 230-95, is actual overcurrent device installed in the circuit breaker or the maximum rating that it can be adjusted to.

Chapter Fifteen: The questions included here were developed using material included in this chapter. The answers can be found by reviewing the text. It is also important that students make use of the 1996 NEC®, where many answers can be found. See page 237 for answers.

1. A system intended to provide protection of equipment from damaging line-to-ground fault currents, by operating to cause a disconnecting means to open all ungrounded conductors of the faulted circuit, is defined as ____.
   a. ground-fault circuit-interrupter
   b. ground-fault protection of equipment
   c. leakage current detector
   d. saturable core reactor

2. The NEC® requires ground-fault protection of all solidly-grounded wye electrical services of more than 150 volts to ground, but not exceeding 600 volts phase-to-phase, for each service disconnect rated ____ amperes or more.
   a. 800
   b. 1,000
   c. 600
   d. 400

3. Ground-fault protection of equipment is not applicable to ____ or continuous industrial processes where a nonorderly shutdown would introduce additional or increased hazards.
   a. electronically-actuated fuses
   b. phase converters
   c. fire pumps
   d. industrial buildings

4. Where a service is protected by a system of GFPE, the maximum setting of the GFPE devices is ____ amperes and the maximum time delay is one second for ground-fault currents equal to or greater than ____.
   a. 1,300 - 2,000
   b. 1,200 - 3,000
   c. 1,400 - 1,500
   d. 1,600 - 1,000

5. For feeders rated 1,000 amperes or more in a solidly-grounded wye system with greater than 150 volts to ground, but not exceeding 600 volts phase-to-phase, ground-fault protection ____ required.
   a. is
   b. is not
   c. sensors are
   d. indication is

6. The most common types of equipment ground-fault protection equipment is:
   a. three-phase, three wire
   b. residual and bypass relaying
   c. zero-sequence and residual
   d. isolation transformer and relay

7. One of the most critical elements of a ground-fault protection system is:
   a. the neutral must be isolated downstream
   b. a choke coil is installed downstream
   c. capacitors are installed downstream
   d. all the above

8. The equipment ground-fault protection system must be tested:
   a. by pushing the "push to test" button
   b. on an annual basis
   c. within 30 days of installation
   d. when first installed on site

9. A second level of equipment ground-fault protection is required
   a. for fire pump installations
   b. for health care facilities
   c. for emergency systems
   d. for optional standby systems

10. Equipment ground-fault protection is required for electrical systems of more than 150 volts to ground and less than 600 volts phase-to-phase:
   a. for feeders of more than 800 amperes
   b. at a building disconnecting means rated at 1200 amperes or more
   c. for feeders rated more than 100 amperes
   d. for a continuous industrial process

Chapter 19

Special Location Grounding and Bonding

Objectives

After studying this chapter, the reader will be able to understand

- Requirements for static and lightning protection.
- Bonding in hazardous (classified) locations.
- Requirements for agricultural buildings.
- Requirements for health care facilities.
- Requirements for swimming pools, hot tubs and spas.
- Requirements for electric signs.

Static protection.

Protection against static electricity is a requirement of a number of industries and establishments. The grounding of equipment is not necessarily a solution to static problems. Each static problem requires its own study and solution. Humidity plays an important part. The higher the humidity the less the chances of static. In some industries, increasing the humidity in the area of a static discharge has been found very effective. One example is in the printing industry.

Lightning protection.

Lightning protection is an important factor in an outdoor substation and at locations where thunder storms are prevalent. Lightning discharges usually consist of very large currents of extremely short duration. Protection is accomplished by deliberately providing a path of low resistance to earth, compared to other paths. There is no guarantee that lightning will necessarily follow the lower resistance path which has been provided, but at least the low resistance path will reduce damage.

The steel fence around an outdoor substation, when properly grounded, offers good lightning protection for the equipment, especially if the fence is higher than the equipment within the station. Lightning conductors will enhance the safety. Lightning rods should project at least one foot above any part of the structure, and equipment, and the path to ground should be as direct as possible.

Each rod should have at least two connections to ground and rods placed on each side of the structure. Surge arresters also should be used on the power lines and the equipment. The lightning conductor system should be connected to earth through driven electrodes as well as being bonded to the station ground bus. Lightning conductors should not be smaller than No. 4, increasing in size as the primary voltage of the system increases.

See NFPA-780, *Lightning Protection Code* for additional information on the subject.

Spacing from lightning rods.

In Section 250-46, the Code requires that metal raceways, enclosures, frames and other noncurrent-carrying metal parts of electric equipment be kept at least 6 feet away from lightning rod conductors.

Where it is not practicable to obtain such 6 feet of separation, the metal enclosure(s) and the lightning rod conductor must be bonded together at the location where the separation distance is less than 6 feet.

Hazardous (classified) locations.

Bonding: The Code places some special requirements for bonding in hazardous (classified) locations. These requirements can be found in Section 501-16 for Class I locations, Section 502-16 for Class II locations and Section 503-16 for Class III locations.

Bonding for Hazardous (Classified) Locations

Class I - 501-16
Class II - 502-16
Class III - 503-16

Figure 16-1

Generally, these requirements are that locknuts on each side of the enclosure, or a locknut on the outside and a bushing on the inside, cannot be used for bonding. Bonding locknuts or bonding bushings with bonding jumpers must be used to ensure the integrity of the bond and its capability of carrying the fault current that may be imposed.

Bonding at Hazardous (Classified) Locations

Bonding required at each point from hazardous location to service equipment

Figure 16-2

The bonding means required here must be installed from the hazardous (classified) location to the service equipment, or point of grounding of a separately derived system that is the source of the circuit. This includes all junction boxes, enclosures, controllers and panelboards in between.

**Exception to Bonding Rules,
Hazardous (Classified) Locations**

Bonding required in this building

Service equipment or separately derived system

Bonding required by main rule does not have to extend back to the service or separately derived system.

Bonding requiremets end at building disconnecting means with branch-circuit overcurrent protection

Figure 16-3

The exception to this section clarifies that the bonding is required to be taken no further than the building disconnecting means that is grounded in accordance with Section 250-24(a), (b) or (c), and that branch-circuit overcurrent protection is located on the load side of the disconnecting means. See Chapter 13 of this text for additional information on grounding electrical systems at additional buildings on the premises.

Grounding in hazardous locations. Where flexible metal conduit or liquidtight flexible metal conduit is permitted and used in Class I, Division 2 hazardous (classified) locations, they must have an internal or external bonding jumper installed to supplement the conduit. If installed outside the conduit, the bonding jumper is limited to 6 feet in length by Section 250-79. See Section 501-16(b).

In these Class I, Division 2 areas, the bonding jumper is permitted to be omitted under the following conditions:

(a) Listed liquidtight flexible metal conduit not more than 6 feet long with fittings listed for grounding is used;

(b) Overcurrent protection of the circuit in the raceway does not exceed 10 amperes; and

(c) The load is not a power utilization load.

Agricultural buildings.

Section 547-8 contains some specific requirements for grounding and bonding in agricultural buildings to which the article applies. The major concern is twofold. The first is for the integrity of the grounding path due to corrosive conditions that exist in these buildings. The second concern is for neutral-to-earth and stray voltage which, if excessive, cause behavior responses in livestock. These behavior responses in dairy cattle can lead to loss of production and health problems.

Section 250-24(a) Exception No. 2 and Section 250-24(b) Exception No. 2 require that the portion of an equipment grounding conductor that is run underground to an agricultural building be insulated or covered copper.

Equipment grounding.

Where equipment that must be grounded is installed in an agricultural building, a copper equipment grounding conductor must be installed to ground the equipment. This applies regardless of the type of wiring method employed. Where installed underground, the equipment grounding conductor must be either insulated or covered copper. See Section 547-8(c).

The frame of water pumps, including the submersible type, must be grounded. [Section 250-42(k)] Where

Equipment Grounding in Agricultural Buildings

- Generally, must comply with Article 250.
- Copper equipment grounding conductor required in wiring method.
- Bond metal well casing for submersible pump to pump circuit equipment grounding conductor.
- Ground motors per Article 430, Part L.

Figure 16-4

submersible pumps are used in metal well casings, the equipment grounding conductor must be bonded to the casing. [Section 250-42(1)].

Isolated neutral or grounded conductor.

One of the most important elements of a farm wiring system, especially where dairy cattle are involved, is to isolate the system neutral at barns, milking parlors, etc. Separation of the neutral conductor from the equipment grounding conductors will prevent voltage drop on the feeder neutral from becoming stray voltage in the building.

Isolation of the neutral is permitted by Section 250-24(a) and (b) Exception No. 2. In the case of agricultural buildings, the conditions of Section 547-8(a) must be satisfied. The neutral-to-ground

Insulated Neutral at Agricultural Building

Feeder

← Service

Distribution → panelboard

Grounding → electrode

Feeder may be overhead or underground. Equipment ground insulated or covered copper if underground.

Main bonding jumper (neutral to enclosure) connection not required at distribution panelboard in or on buildings housing livestock or poultry where all the following conditions are met:
  a. All buildings and premises wiring are under the same ownership.
  b. An equipment grounding conductor is run with the supply conductors and is of the same size as the largest supply conductor, if of the same material, or is adjusted in size in accordance with the equivalent size columns of Table 250-95 if of different materials.
  c. Service disconnecting means is provided at the distribution point for supply to those buildings.
  d. The equipment grounding conductor is bonded to the grounded circuit conductor at the service equipment or the source of a separately derived system.
  e. A grounding electrode is provided and connected to the equipment grounding conductor in the distribution panelboard.

Figure 16-5

bonding jumper is not required at the distribution panelboard in or on buildings housing livestock or poultry where all the following conditions are complied with:

a. All buildings and premises wiring are under the same ownership.

b. An equipment grounding conductor is run with the supply conductors and is of the same size as the largest supply conductor, if of the same material, or is adjusted in size in accordance with the equivalent size columns of Table 250-95 if of different materials.

c. Service disconnecting means is provided at the distribution point for supply to those buildings.

d. The equipment grounding conductor is bonded to the grounded circuit conductor at the service equipment or the source of a separately derived system.

e. A grounding electrode is provided and connected to the equipment grounding conductor in the distribution panelboard.

Where this provision is employed, service equipment is often installed at a location remote from the agricultural building, and feeders are run to the agricultural buildings. Service switches are installed at the metering point, often referred to as a "May Pole" as feeders radiate from the service switch location to the various buildings. The neutral and equipment grounding conductor are bonded together in the service equipment.

It is not uncommon to find one or more transfer switches at this location for connection of portable or tractor-mounted generators for standby power.

Stray (Tingle) voltage. Voltage drop on a neutral conductor supplying a building housing livestock can result in elevated levels of stray

Equipotential Bonding in Agricultural Buildings

Panelboard

Rebar, mesh or similar.
Copper bonding conductor not smaller than No. 8 (typical).

Figure 16-6

voltage. It is important to balance 120-volt loads to minimize neutral loads, operate motors at 240 volts wherever practical and size neutral conductors in service drops and feeders as large as practical.

Livestock behavioral responses, including production and health problems, can be caused by stray voltage (also referred to as "tingle-voltages"). These voltages appear between various portions of grounded or ungrounded metallic systems such as electrical equipment or piping systems and the earth or floor such that livestock can come between two different potentials. It is not uncommon to find voltage differences between adjacent concrete slabs or between a concrete slab and the adjacent earth.

Common causes of neutral-to-earth voltages (stay voltage if it is at livestock contact points) are currents on primary distribution systems, farm secondary neutral conductors and faulty wiring on the farm. Some electric utilities have installed primary-to-secondary neutral conductor isolators in an attempt to solve the stray voltage problem. Neutral conductors that are too small for the load and length of run and loose or corroded splices and terminations will frequently elevate the neutral-to-earth voltage to a level that can affect livestock.

Another common cause of stray voltage is ground faults where the fault current gets into the earth, concrete slabs or on metal equipment that may be contacted by livestock. Ground faults that can cause stray voltage may occur in water pumps, underground wires, sump pumps, manure pumps, electrically heated livestock watering fountains, electrically operated feeders and similar equipment. Proper grounding along with bonding this equipment together usually prevents the stray voltage from occurring even where there is a high impedance fault.

Equipotential plane. Section 547-8(b) requires the installation of an equipotential plane in the concrete floor of livestock confinement areas. This is the best method of controlling the effects of stray voltages on livestock. This bonding does not correct or remove the faults that are causing the problem but keeps everything in the livestock area at the same potential. This, in essence, prevents the livestock from being aware of and being affected by the stray voltage.

Voltage Gradient at Entrances and Exits

Pipe framework

Livestock entrance or exit

Gradient ramp

Conductive Element = - - - - - - -
Approved Connection = •

8-10 ft. Ground rods
45° angle
2 or more per door

Figure 16-7

The term, "equipotential plane" is defined in Section 547-8 as, "An equipotential plane is an area where a wire mesh or other conductive elements are embedded in concrete, bonded to all adjacent conductive equipment, structures, or surfaces, and connected to the electrical grounding system to prevent a difference in voltage from developing within the plane."*

Wire mesh or other conductive elements, such as reinforcing steel rods, must be installed in these areas and be bonded together, as well as to the building grounding electrode system. This is accomplished by bonding the reinforcing rods and wire mesh in the concrete together with a copper conductor, not smaller than No. 8, and then connecting these bonding conductors to metal piping systems, stanchions and grounding electrode systems. This creates the required equipotential plane which really means that everything in the area is at the same potential. This prevents voltage differences between conductive bodies livestock can make contact with.

Obviously, it is best to install this bonding system during the original construction of the building or portion of a building that houses livestock or serves as a milking parlor. Equipotential planes have been installed in existing buildings by sawing grooves in the concrete floors, installing bonding conductors and bonding all conductive elements together.

It is also required that there be a voltage gradient at livestock entrances and exits. This is accomplished as indicated by a Fine print Note that follows Section 547-8(b). It reads, "One method to establish a voltage gradient is in accordance with Equipotential Planes in Animal Containment Areas, American Society of Agricultural Engineers (ASAE) EP473-1994."*

The method indicated for producing a voltage gradient is installing two or more 8- to 10-foot long ground rods at each livestock exit and entrance. The ground rods are installed at a 45 degree angle with the highest end at the building. The rods are connected by means of a bonding conductor to the equipotential plane of the building.

Bonding isolator. A listed impedance device, that essentially consists of a saturable core reactor, is permitted by Section 547-8 Exception No. 2 to be used to bond interior metal piping systems in agricultural buildings. This device can be used only where electrical equipment requiring grounding is not attached to, or in contact with, the metal piping systems. The device is "open" electrically as long as there is little or no voltage across the device. It "closes" and conducts current when a predetermined voltage appears across the device.

Bonding Isolator in Agricultural Buildings

Figure 16-8

At best, the grounding isolator is a "band-aid" approach for existing installations. Bonding conductive portions together, including the floor and piping systems, will almost always be more effective in solving the stray-voltage problem.

Health care facilities.

Electrical systems in health care facilities must comply with two safety standards, the *National Electrical Code®*, NFPA 70 and the *Health Care Facilities Code®* NFPA-99. These codes provide the minimum requirements for installation as well as maintenance and testing of electrical systems in health care facilities.

Grounding in patient care areas.

Section 517-13(a) generally requires that, "In an area used for patient care, the grounding terminals of all receptacles and all noncurrent-carrying conductive surfaces of fixed electric equipment likely to become energized that are subject to personal contact, operating at over 100 volts, shall be grounded by an insulated copper conductor. The grounding conductor shall be sized in accordance with Table 250-95 and installed in metal raceways with the branch-circuit conductors supplying these receptacles or fixed equipment."*

Exception No. 1 provides that, "Metal raceways shall not be required where listed Types MI, MC, or AC cables are used, provided the outer metal armor or sheath of the cable is identified as an acceptable grounding return path."

Exception No. 2 provides that, "Metal faceplates shall be permitted to be grounded by means of a metal mounting screw(s) securing the faceplate to a grounded outlet box or grounded wiring device."

Grounding of Lighting Fixtures in Patient Care Area

Figure 16-9

Finally, Exception No. 3 provides that, "Light fixtures more than 7-½ feet (2.29 m) above the floor shall not be required to be grounded by an insulated grounding conductor."

Both a primary and secondary, redundant means of grounding is required in patient care areas of health care facilities. See Section 517-13. The metal raceway or outer metal jacket, that is suitable for grounding, of a listed cable provides the primary means of equipment grounding. An insulated copper equipment grounding conductor (the secondary, redundant equipment grounding means) must also be installed for grounding receptacles and noncurrent-carrying metal portions of fixed electric equipment in the patient care areas that are subject to personal contact.

The wiring methods that are acceptable for providing the primary equipment grounding fault return path are the metal raceways or cable assembly included in Section 250-91(b). (See the following table.) Of course, where these wiring methods are installed, it is critical that good work practices be followed to ensure that the wiring method provides a reliable, continuous and low impedance path for fault current.

Health Care Facilities
Patient Care Area Wiring Methods

1. **Rigid metal conduit.**
2. **Intermediate metal conduit.**
3. **Electrical metallic tubing.**
4. **Listed flexible metal conduit, max. 6 feet long, max. 20A OC protection, fittings listed for grounding.**
5. **Armor of listed Type AC cable where the outer metal jacket is approved as a grounding means.**
6. **Listed Type MI cable and MC cable where the outer metal jacket is acceptable as a ground-return path.**
7. **Listed liquidtight flexible metal conduit maximum 6 feet long, and Max. 20 A OC protection ³⁄₈ to ½ inch or Max. 60 A OC protection ¾ - 1¼ inch**

All wiring methods must contain an insulated, copper equipment grounding conductor.

Sections 517-13(a) and (b) place additional restrictions on the use of cable wiring methods that are used for wiring in patient care areas. The outer jacket of the cable is required to qualify as an equipment grounding return path in accordance with Section 250-91(b).

Types MC and MI cables are required in Section 517-13(b) to have an outer armor that is "identified" (defined in Article 100) "as an acceptable

grounding return path." These requirements do not permit the combination of the cable outer jacket, which alone does not qualify as an equipment grounding conductor and an internal equipment grounding conductor, to meet the requirements for grounding.

Type AC Health Care Facilities Cable

1. Galvanized steel armor.
2. Insulated bushing.
3. Green insulated ground conductor.
4. Nylon.
5. Thermoplastic insulation 90°C rated.
6. Copper conductors.
7. Bonding wire.

Figure 16-10

By its construction and listing by a qualified electrical testing laboratory, Type AC cable with a spiral interlocking metal jacket and a bonding strip or wire in intimate contact with the outer jacket is suitable as an equipment grounding conductor.

Generally, the outer jacket of Type MC cable of the spiral-interlocking type is not suitable as an equipment grounding conductor and cannot be used in a patient care area of a health care facility that is required to comply with Section 517-13. Type MC cable with a smooth or corrugated continuous tube is suitable for these patient care areas if it contains a green-insulated equipment grounding conductor with a yellow stripe or surface marking or both to indicate that it is an additional equipment (or isolated) grounding conductor. This acceptable type of MC cable would not contain a supplemental bare or unstriped green-insulated grounding conductor. A supplemental bare or unstriped green-insulated equipment grounding conductor in Type MC cable of the smooth or corrugated continuous type is an indication that the outer armor by itself does not qualify as an equipment ground.

Copper, Type MI cable, by nature of its construction, has an outer jacket that is suitable as an

Flexible Metal Conduit and Liquidtight Flexible Metal Conduit in Patient Care Areas

Metal raceway with EGC

Listed liquidtight flexible
metal conduit.
Maximum 6 ft. long.
Maximum 20 amps OC device
3/8" and 1/2".
Maximum 60 amp 3/4" - 1¼".
Insulated equipment grounding wire.

Listed flexible metal conduit.
Maximum 6 ft. long.
Maximum 20 amps OC device.
Fittings listed for grounding.
Insulated equipment grounding wire.

Figure 16-11

equipment grounding conductor. This cable is manufactured with bare conductors that are physically separated from each other by the mineral compound inside the metal sheath. The individual conductors are insulated by slip-on insulation at the time the cable is terminated.

While recognized in Section 250-91(b) as equipment grounding conductors, severe restrictions are placed on the use of flexible metal conduit and liquidtight flexible metal conduit due to their limited capabilities of carrying current and functioning as a ground-fault return path. These restrictions are illustrated in Figure 16-11.

Additional restrictions are placed on the wiring of the emergency system in hospitals. While these requirements relate to mechanical protection of the circuits and not directly to grounding, the installer should be aware of these rules. See Section 517-30(b)(2) for designation of these branches, Section 517-3 for definitions and Section 517-30(c)(3) for wiring requirements.

Testing of grounding system in patient care areas.

The *Standard for Health Care Facilities*®, NFPA 99-1996, requires that the integrity of the grounding path provided by the wiring method in patient care areas of health care facilities be tested before acceptance of the initial installation and after any alterations or replacement of the electrical system is made. See Section 3-3.3.2.1. Both voltage and impedance measurements must be made of ex-

posed conductive surfaces including the grounding contacts of receptacles in the patient-care vicinity.

Excluded from the testing requirements are small, wall-mounted conductive surfaces, not likely to become energized, such as surface-mounted towel and soap dispensers, mirrors and the like. Also exempt are large, metal conductive surfaces not likely to become energized, such as window frames door frames and drains. A note following this section indicates that the grounding system, including both the metallic raceway and equipment grounding conductor, is to be tested as an integral system. Removing equipment grounds from receptacles or equipment is not required or recommended.

Testing Grounding System Patient Care Area

Figure 16-12

The voltage and impedance measurements are required to be measured against a reference point. The reference point is permitted to be one of the following:

1. a reference grounding point,
2. a grounding point in the room under test that is electrically remote from the equipment under test, such as a metal water pipe,
3. the grounding contact of a receptacle that is powered from a different branch circuit from the receptacle being tested.

The criteria that must be met are as follows:

1. the voltage limit is 20 mV.
2. the impedance limit is 0.1 ohm.
3. for "quiet ground systems" the impedance limit is 0.2 ohm.

Receptacle testing in patient care areas.

Unless documented performance to substantiate testing at longer intervals is provided, receptacles in patient care areas must be tested for grounding reliability at least every:

1. 12 months for general care areas.
2. 6 months for critical care areas.
3. 12 months for wet locations.

Tests that must be performed for each receptacle include:

1. visual inspection to confirm physical integrity.
2. continuity of the grounding circuit.
3. correct polarity.
4. retention force (except locking type) to be not less than 115 g (4 oz).

Where ground-fault circuit-interrupters are installed, they must be tested for operation not less than every 12 months.

Bonding of panelboards in patient care areas.

Section 517-14 requires that the equipment grounding terminal busses of the normal and essential branch-circuit panelboards be bonded together. The bonding conductor must be insulated continuous copper not smaller than No. 10. Where more than two panelboards serve the same location, it is permitted to install the bonding conductor unbroken from one panelboard equipment grounding terminal bar to another without looping an unbroken conductor between all panelboard enclosures.

Panelboard Bonding, Patient Care Areas

Insulated, continuous copper not smaller than No. 10

Figure 16-13

This bonding is intended to ensure that little if any potential difference exists between exposed noncurrent-carrying metal portions of equipment in the patient-care area. It is not required that this bonding conductor be installed in a raceway between panelboards where it is protected by normal building construction such as studs and gypsum board.

Patient equipment grounding point.

The "Patient Equipment Grounding Point" is an option for improving electrical safety in the patient care area. Though an optional feature [Section 517-19(c)], a patient equipment grounding point is provided by many hospital equipment manufacturers and specified by consulting engineers. Where installed, it consists of a grounding terminal bus and may include one or more grounding jacks. An equipment bonding jumper not smaller than No. 10 is then used to bond the grounding terminals of all receptacles in the patient vicinity to the equipment grounding point. Again, this is done to reduce the potential difference between conductive surfaces in the patient vicinity.

Panelboard grounding, critical care areas.

Section 517-19(d) requires that where metal raceways or Type MC or MI cables are used for feeders, grounding of panelboards and switchboards serving critical care areas bonding must be assured. Acceptable methods include as follows:

Panelboard Grounding, Critical Care

Grounding bushing with bonding jumpers

Threaded bosses or hubs on enclosures

Bonding locknuts or bushings

Figure 16-14

1. a grounding bushing and a continuous copper bonding jumper sized in accordance with Section 250-95, with the bonding jumper connected to the junction enclosure or the ground bus of the panel, or
2. connection to threaded bosses or hubs on enclosures, or
3. other approved devices such as bonding locknuts or bushings.

Swimming pools, general.

It is important to understand the organization of Article 680 for proper application of the requirements. Article 680 is divided in several parts, namely Part A through Part G. Part A is general and applies to all equipment and installations covered in Article 680. Part B covers permanently installed pools. Other parts cover the equipment or installation identified in the title of the part and do not necessarily apply to all equipment included in Article 680. For example, the rules in Part B for grounding panelboards do not apply to the wiring of a hydromassage bathtub covered in Part G.

However, be aware of the requirements for wiring spas and hot tubs. Section 680-40 tells us that "A spa or hot tub installed outdoors shall comply with the provisions of Parts A and B." Four exceptions follow. Generally, however, a spa or hot tub installed outdoors must meet all the rules for swimming pools, while a spa or hot tub installed indoors must meet the rules in Section 680-41 which are different from those for swimming pools.

Keep this distinction in mind as we review several of the grounding and bonding requirements. Basically, the requirements from Section 680-20 through 680-28 apply to both permanently installed pools and spas and hot tubs that are installed outside.

GFCI requirements for swimming pools.

Sections 680-5(b) and (c) contain specific requirements relative to ground-fault circuit-interrupters for swimming pools. They are permitted to be self-contained units, circuit-breaker types, receptacle types, or other approved types.

Conductors on the load side of a ground-fault circuit-interrupter used to provide protection for underwater lighting fixtures are generally not permitted to occupy raceways, boxes or enclosures containing other conductors, even if the other conductors are for pool-related equipment such as for a water pump.

Four exceptions to the general rule are given. They are as follows:

Exception No. 1: Ground-fault circuit-interrupters shall be permitted in a panelboard that contains circuits protected by other than ground-fault circuit-interrupters.

Exception No. 2: Supply conductors to a feed-through-type ground-fault circuit-interrupter shall be permitted in the same enclosure.

Exception No. 3: Conductors on the load side of a ground-fault circuit-interrupter shall be permitted to occupy raceways, boxes, or enclosures containing only conductors protected by ground-fault circuit-interrupters.

Exception No. 4: Grounding conductors.

Wiring For Underwater Lights

GFCI protected conductors permitted in panelboard

To pool pump or other non-GFCI protected circuit

To underwater light(s). Only GFCI protected conductors permitted

Supply conductors permitted to GFCI device, not in raceway on load side of GFCI if conductors supply underwater light

Figure 16-15

Table 16-1
GFCI Requirements in Article 680

Receptacles, pools	680-6(a)(3)
Lighting fixtures, ceiling	
Fans, pools	680-6(b)(1)
	680-6(b)(2)
Underwater lighting fixt.	680-20(a)(1)
Pool cover	680-26(b)
Storable pool equipment	680-31
Receptacles, spa & hot tub	680-41(a)(2)
Lighting fixtures, ceiling	
fans, spa & hot tub	680-41(b)(1)
Spa or hot tub, outlet for	680-42
fountains	680-51(a)
	680-56(a)
Therapeutic pools and tubs	680-62(a)
Hydromassage bathtubs	680-70

GFCI requirements for receptacles.

Section 680-6(a) generally requires that receptacles on the property be located at least 10 feet from the inside walls of a pool (outside spa or hot tub) or fountain. The exception permits a receptacle(s) that provides power for a water-pump for a permanently installed pool or fountain, as permitted in Section 680-7, to be installed between 5 and 10 feet from the inside walls of the pool or fountain. Where so located, the receptacle(s) must be of the single and of the locking and grounding types and must be protected by a GFCI device.

GFCI Requirements for Receptacles

Pool or outdoor spa or hot tub

GFCI required

GFCI not required

8'

GFCI not required

Sliding door

Window that opens

GFCI not required

Permitted closer than 10 feet if for permanently installed pool water pump motor. GFCI protection required.

Figure 16-16

Where a permanently installed pool is installed at a dwelling unit(s), at least one 125-volt receptacle is required to be located a minimum of 10 feet from and not more than 20 feet from the inside wall of the pool. This receptacle must be located not more than 6 foot 6 inches above the floor, platform, or grade level serving the pool. (Remember that where the word "pool" is used, it also includes outdoor spas and hot tubs.)

All 125-volt receptacles located within 20 feet of the inside walls of a pool or fountain must be protected by a ground-fault circuit-interrupter.

A fine print note that follows this section provides information on determining the distances from the pool. It reads, "In determining the above dimensions, the distance to be measured is the shortest path the supply cord of an appliance connected to the receptacle would follow without piercing a floor, wall, ceiling, doorway with hinged or sliding door, window opening, or other effective permanent barrier."

Bonding at swimming pools.

Swimming pools, spas and hot tubs present special shock hazards to people due to mixing water enriched with chemicals, people and electricity together. Measures as prescribed in Article 680 of the Code, if carefully followed, will reduce the electric shock hazards to an acceptable level. One of these measures is the bonding together of conductive portions of the pool and metal parts of electrical equipment associated with the pool. The goal is to provide a means of equalizing the potential of all equipment and parts so there will be no current flow between parts. This is accomplished by connecting all the parts together by an adequately-sized and properly-connected copper conductor. This concept is often called equipotential bonding or grounding.

Grounding and Bonding

Approved pool junction box, supported properly

Wet-niche fixture

Equipment Grounding

Bonding, Min. No. 8 CU

Figure 16-17

A Fine Print Note in this section of the Code clarifies that the bonding conductor used for equipotential bonding of equipment in the pool, hot tub or spa vicinity is not required to be extended to a remote panelboard or service equipment or grounding electrode. Since several pieces of electrical equipment commonly used with pools, hot tubs or spas that are bonded together must also be grounded, an interconnection between the grounded (neutral)

Bonding Requirements for Swimming Pools and Spas and Hot Tubs Installed Outdoors

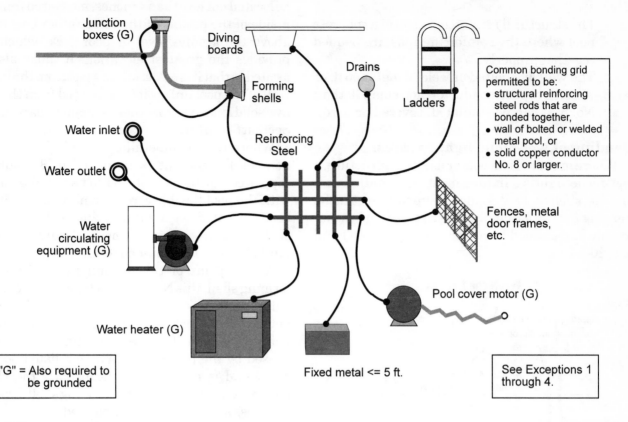

Figure 16-18

conductor and bonding grid exists. This interconnection is often remote and is not intended to play a part in equipotential bonding.

These requirements emphasize the purpose of bonding in a swimming pool, spa or hot tub area - that of equalizing the potential (voltage) between various parts of the pool. By keeping the potential difference as low as practicable, the shock hazard is reduced significantly.

Section 680-22 provides the basic bonding requirements where a list of items to be bonded is given. Included are the following:

1. All metal parts of the pool structure, including the reinforcing metal of the pool shell, coping stones and deck,
2. All metal forming shells and mounting brackets of no-niche fixtures.
3. All metal fittings within or attached to the pool structure. (Isolated parts that are not over 4 inches in any dimension and do not penetrate into the pool structure more than 1 inch are exempted.)
4. Metal parts of electric equipment associated with the pool water circulating system, including pump motors.

5. Metal parts of equipment associated with pool covers, including electric motors.
6. Metal-sheathed cables and raceways, metal piping, and all fixed metal parts that are within 5 feet horizontally of the inside walls of the pool, and within 12 feet above the maximum water level of the pool, or any observation stands, towers or platforms, or from any diving structures, that are not separated from the pool by a permanent barrier.

Common bonding grid.

These parts are to be connected to a common bonding grid with a solid copper conductor not smaller than No. 8. The conductor can be insulated, covered or bare. Connections to the parts to be bonded and to the common bonding grid must be made by pressure connectors or clamps of stainless steel, brass, copper or copper alloy. (It is interesting to note here that exothermic welding, as allowed by Section 250-113, for making grounding and bonding connections is not recognized for similar purposes in Article 680.)

An exception recognizes steel tie wires as being suitable for connecting the reinforcing steel together. The steel tie wires must be made up tightly.

Acceptable common bonding grids include as follows:

1. The structural reinforcing steel of a concrete pool where the reinforcing rods are bonded together properly.
2. The wall of a bolted or welded metal pool, or
3. A solid copper conductor not smaller than No. 8. It can be insulated, covered or bare.

Bonding of wet-niche light fixtures.

Where mounted in a pool or fountain structure, a wet-niche lighting fixture must be installed in a forming shell that is designed to support a wet-niche lighting fixture assembly. The fixture will be completely surrounded by water. [Sections 680-3 and 680-20(b)]

Wet-Niche Fixtures

Approved pool junction box, supported properly

4 ft. or more

Min. 4" above deck and 8" above water

Metal conduit of brass or equal

Wet-niche fixture

If PVC or liquidtight flexible nonmetallic conduit is used, install No. 8 insulated copper conductor. Encapsulate connection in forming shell with listed potting compound.

Rigid metal, intermediate metal or rigid nonmetallic conduit. Section 680-25(b)(1)

Figure 16-19

A forming shell must be installed for the mounting of all wet-niche underwater lighting fixtures. The forming shell must also be equipped with provisions for threaded conduit entries.

Conduit permitted to be used to connect the forming shell to a suitable junction box or other permitted enclosure include rigid metal, intermediate metal liquidtight flexible metallic or rigid nonmetallic. [Section 680-20(b)(1)]

Rigid metal or intermediate metal conduit used to connect the wet-niche fixture housing must be made of brass or other approved corrosion resistant metal, such as stainless steel.

The junction box or other suitable enclosure must be located not less than 4 inches, measured from the inside of the bottom of the box or other enclosure, above the ground level or pool deck, whichever provides the greatest elevation. It must also be located so that it is not less than 4 feet from the inside wall of the pool unless it is separated from the pool by a solid fence, wall or other permanent barrier. See Section 680-21.

Where rigid nonmetallic or liquidtight flexible nonmetallic conduit is used, a No. 8 insulated copper conductor is required to be installed along with the pool flexible cord assembly so that it can be terminated on a suitable lug in the forming shell, junction box, transformer enclosure or ground-fault circuit-interrupter enclosure.

At the point of this termination, within the forming shell, this No. 8 conductor must be covered with, or encapsulated in, a listed potting compound. [Section 680-20(b)(1)] This compound provides protection from deteriorating effects often caused by the pool water. Where a listed potting compound is not used to encapsulate the No. 8 bonding conductor inside the forming shell, brass or stainless steel conduit must be used.

Where in contact with the pool water, metal parts of the fixture and forming shell must also be of brass or other approved corrosion-resistant metal.

The end of the flexible cord jacket and the flexible cord conductor terminations located within a wet-niche fixture are required to be covered with, or encapsulated in, a suitable potting compound. [Section 680-20(b)(2)] This will help to prevent the siphoning of water into the fixture through the cord jacket or its contained conductors.

In addition, an equipment grounding conductor connection within a wet-niche fixture must be similarly treated to protect it from the deteriorating effect of pool water in the event of water entry into the fixture.

The fixture must be bonded to and secured to the forming shell by a positive locking device that will ensure a low resistance contact. A special tool is required in order to remove the fixture from the forming shell.

Underwater wet-niche lighting fixtures include markings that will indicate the proper housing or housings with which they are to be used, and the fixture housings are marked to indicate the fixture or fixtures with which the housings are to be used. These fixtures are provided with a factory-installed

Grounding of Metal Electrical Parts

Legend:
1. Insulated copper equipment grounding conductor sized in accordance with Table 250-95 but not smaller than No. 12.
2. Install in rigid metal conduit, intermediate metal conduit or rigid nonmetallic conduit.
3. Install in rigid metal conduit, intermediate metal conduit or rigid nonmetallic conduit or Type MC cable listed for application.
4. Electrical metallic tubing permitted inside buildings.
5. Electrical nonmetallic tubing permitted inside buildings per Article 331
6. Liquidtight flexible metal or nonmetallic conduit permitted where flexibility required.
7. Wiring method of Chapter 3 with equipment grounding conductor insulated or covered by outer sheath not smaller than No. 12 permitted in one-family dwelling.
8. Flexible cord permitted per Section 680-7.
9. Wet-niche or no-niche fixture supplied by flexible cord; insulated copper conductor not smaller than circuit conductors, not smaller than No. 16.

Figure 16-20

permanently-attached flexible cord that extends at least 12 feet outside the fixture enclosure. This will permit removal of the fixture from the forming shell so that it can be lifted onto the pool or spa deck for servicing without lowering the water level or disconnecting the fixture from the branch circuit conductors.

Fixtures with longer cords are available for installation where the junction box or splice enclosure is so located that a 12-foot-long cord will not permit the fixture to be removed from the forming shell and placement on the deck for servicing. To avoid possible cord damage, any cord length in excess of that necessary for servicing should be trimmed from the supply end rather than stored in the forming shell.

Grounding requirements.

The following swimming pool and outdoor spa and hot tub equipment is required to be grounded:
1. Wet-niche underwater lighting fixtures;
2. Dry-niche underwater lighting fixtures;
3. All electric equipment located within 5 feet of the inside wall of the pool;
4. All electric equipment associated with the recirculating system of the pool;
5. Metal junction boxes;
6. Metal transformer enclosures;
7. Ground-fault circuit-interrupters;
8. Panelboards that are not part of the service equipment and that supply any electric equipment associated with the pool. See Section 680-24.

Methods of grounding.

Wet-niche, dry-niche and no-niche lighting fixtures must be connected to an equipment grounding conductor that is sized in accordance with Table 250-95. This equipment grounding conductor must be an insulated copper conductor and can never be smaller than No. 12. It must be installed with the circuit conductors in a rigid metal conduit, intermediate metal conduit, or rigid nonmetallic conduit. [Section 680-25(b)(1)]

Electrical metallic tubing, and electrical nonmetallic tubing installed in compliance with Article

331, are permitted to be used to protect circuit conductors where installed on or within buildings.

The equipment grounding conductor installed between the wiring chamber of the secondary winding of a transformer and a junction box must be sized in accordance with the overcurrent device in this circuit. See Table 250-95.

The junction box, transformer enclosure, or other enclosure in the supply circuit that is run to a wet-niche lighting fixture, and the field wiring chamber of a dry-niche lighting fixture, must be grounded to the equipment grounding terminal of a metal panelboard. This terminal must be directly connected to the metal panelboard enclosure, and the equipment grounding conductor must be installed without any joint or splice.

Where more than one underwater lighting fixture is supplied by the same branch circuit, the equipment grounding conductor installed between the junction boxes, transformer enclosures, or other enclosures in the supply circuit to wet-niche fixtures, or between the field wiring compartments of dry-niche fixtures, are permitted to be terminated on equipment grounding conductor terminals.

Where the underwater lighting fixture is supplied from a transformer, ground-fault circuit interrupter, clock-operated switch, or a manual snap switch that is located between the panelboard and a junction box connected to the conduit that extends directly to the underwater lighting fixture, the equipment grounding conductor is permitted to terminate on equipment grounding terminals on the transformer, ground-fault circuit-interrupter, clock-operated switch enclosure, or an outlet box that is used to enclose a snap switch.

Wet-niche lighting fixtures that are supplied by flexible cord or cable assemblies must have all their exposed noncurrent-carrying metal parts grounded by an insulated copper equipment grounding conductor that is an integral part of the cord or cable. This equipment grounding conductor must be connected to an equipment grounding terminal in the supply junction box, transformer enclosure or other enclosure. Also, this equipment grounding conductor cannot be smaller than the supply conductors and never smaller than No. 16.

Pool associated motors.

All pool associated motors are required to be connected to an equipment grounding conductor sized in accordance with Table 250-95. [Section 680-25(c)] This equipment grounding conductor must be an insulated copper conductor and can never be smaller than No. 12. This equipment grounding conductor must be installed with the circuit conductors in a rigid metal conduit, intermediate metal conduit, rigid nonmetallic conduit or Type MC metal-clad cable specifically listed for the application.

Electrical metallic tubing is permitted to be used to protect conductors where it is installed on or within buildings. (Section 680-25(c) Exception No. 1)

Where flexible connections are necessary at or adjacent to the motor, liquidtight flexible metal or liquidtight flexible nonmetallic conduit with approved fittings are permitted. (Section 680-25(c) Exception No. 2)

Pool Associated Wiring, One-Family Dwelling Unit Exception

Any of the wiring methods recognized in Chapter 3 of the NEC® that contain an equipment grounding conductor that is insulated or covered by the outer sheath of the wiring method and is not smaller than No. 12 is permitted in the interior of one-family dwellings to supply pool associated motors.

Figure 16-21

For pump motors only, any of the wiring methods recognized in NEC® Chapter 3 containing an equipment grounding conductor that is either insulated or covered by the outer sheath of the wiring method, and sized to be not smaller than No. 12, are permitted to be used for the wiring method inside of a one-family dwelling. These wiring methods include Type NM cable and Type UF cable. (Section 680-25(c) Exception No. 3).

Flexible cord is permitted to be installed in accordance with Section 680-7. (Section 680-25(c) Exception No. 4)

Permanently installed pools are permitted to be provided with listed cord- and plug-connected pool pumps that are protected by a system of double insulation. They are to be provided with a means for grounding only the internal and nonaccessible non-current-carrying metal parts of the pump. (Section 680-28)

Panelboard Grounding

Service equipment

Bonded equipment grounding terminal bar

Subpanel

Bonded neutral

Isolated neutral

Feeder with insulated equipment grounding conductor installed in rigid metal conduit, intermediate metal conduit or rigid nonmetallic conduit. (See exception)

Ex. No. 1 - Flexible metal conduit or approved cable with insulated or covered equipment ground permitted for existing remote panelboard.
Ex. No. 2 - EMT permitted on or within building.

Figure 16-22

Panelboard grounding.

A panelboard that is not part of the service equipment or source of a separately derived system is required to have an equipment grounding conductor installed between its equipment grounding terminal, and the equipment grounding terminal located at the service equipment or source of a separately derived system.

This conductor must be sized in accordance with Table 250-95 and must be an insulated conductor of copper, aluminum or copper-clad aluminum and can never be smaller than No. 12. This conductor must be installed with the feeder conductors in a rigid metal conduit, intermediate metal conduit or rigid nonmetallic conduit. The equipment grounding conductor must be connected to an equipment grounding terminal of the panelboard.

Panelboards from separately derived systems.

There are several ways of grounding swimming pool panelboards that are supplied from a separately derived system. It is important to follow the rules for grounding the separately derived system as given in Section 250-26 and as covered extensively in Chapter 12 of this text.

Section 680-25(c) requires that a panelboard that is not part of the source of a separately derived system must have an insulated equipment grounding conductor installed with the feeder to the swimming pool panelboard. This means that a panelboard that "is" a part of the source of the separately derived system is not required to have an insulated equipment grounding conductor in the supply to the panelboard. There still must be a proper path for fault current return to the source, again as emphasized in Chapter 12 of this text. An equipment grounding conductor must be installed where a panelboard for swimming pool equipment is installed that is downstream from the source of the separately derived system.

Existing panelboards.

The equipment grounding conductor installed between an existing remote panelboard and the service equipment is not required to be run in rigid metal conduit, intermediate metal conduit or rigid nonmetallic conduit, if the interconnection is by means of a flexible metal conduit, or an approved

Swimming Pool Panelboard From Separately Derived System

A panelboard, not part of the source of a separately derived system is required to have an equipment grounding conductor in the supply to the panelboard.
See also Sections 250-26 and 250-61(a)(3).

Figure 16-23

cable assembly provided with an insulated or covered copper, aluminum or copper-clad equipment grounding conductor. (Section 680-25(d) Exception No. 1)

The NEC® does not define what is meant by "existing." Often, inspection authorities consider an installation to be "existing" after it has had final approval or after a certificate of occupancy has been issued.

Electrical metallic tubing is permitted to be used to protect conductors where it is installed on or within the building. (Section 680-25(d) Exception No. 2)

Swimming pool panelboards in other buildings.

Grounding at the disconnecting means for remote buildings is covered in Section 250-24 and in Chapter Thirteen of this text where we give the methods for complying with the Code rules. Normally, one can choose between grounding the neutral conductor again at the building and not taking the equipment grounding conductor to the building, and taking the equipment grounding conductor to the building, grounding it there and not grounding the neutral but "floating" it there.

The second method is the one that must be followed for the disconnecting means at remote buildings that house swimming pool equipment. The equipment grounding conductor that is run with the feeder must be an insulated conductor and must be connected "to the equipment grounding terminal of the panelboard." See Section 680-25(d). This requirement prevents connecting the neutral and equipment grounding conductor together in parallel as is permitted in Section 250-24 for other buildings. Connecting a different size neutral and equipment grounding conductor together in parallel is a violation of Section 310-4(3).

Cord-connected equipment.

Where fixed or stationary equipment is connected with a flexible cord to facilitate removal or disconnection for maintenance, repair, or storage as provided in Section 680-7, the equipment grounding conductor must be connected to a fixed metal part of the assembly. The removable part must be mounted on or bonded to the fixed metal part. [Section 680-25(e)]

All other electrical equipment must be grounded in accordance with Article 250 and must be con-

Grounding for Swimming Pool at Second Building

Feeder — Service — Subpanel →

Feeder may be overhead or underground. Neutral conductor must be isolated from enclosure at second building. Equipment grounding conductor is grounded at second building.

Figure 16-24

nected by an approved wiring method covered in NEC® Chapter 3. See Section 680-25(f).

Spa or hot tub installed outdoors.

A spa or hot tub installed outdoors must comply with the provisions of Article 680, Part A, General Requirements and Part B covering Permanently Installed Pools. (Section 680-40) Four exceptions from the general rule apply.

Metal bands or hoops that are used to secure wooden staves are not required to be bonded to a common bonding grid. (Exception No. 1)

Listed packaged units are permitted to be cord- and plug-connected with a cord not longer than 15 feet, when protected by a ground-fault circuit-interrupter. (Exception No. 2)

The required bonding is accomplished by metal to metal contact where mounted on a common frame or base. (Exception No. 3)

Spa or hot tub installed indoors.

A spa or hot tub installed indoors must conform to the requirements of Article 680, Part D and are required to be connected by wiring methods of NEC® Chapter 3.

Where rated at 20 amperes or less, listed packaged units are permitted to be cord- and plug-connected to facilitate their removal or disconnection for maintenance and repair.

At least one receptacle is required to be located at not less than 5 feet from and not more than 10 feet from the inside wall of a spa or hot tub.

All receptacles on the property must be located

to be at least 5 feet away from the inside walls of the spa or hot tub.

Any 125 volt-rated receptacles that are located within 10 feet of the inside walls of a spa or hot tub must be protected by a ground-fault circuit-interrupter.

The distance measured is the shortest path the supply cord of an appliance connected to the receptacle would follow without piercing a floor, wall or ceiling of a building or other effective permanent barrier.

Receptacles that provide power for a spa or hot tub must also be protected by a ground-fault circuit-interrupter.

Bonding of spa or hot tub installed indoors.

The following parts of a spa or hot tub are required to be bonded together:

1. All metal fittings within or attached to the spa or hot tub structure.

2. Metal parts of any electric equipment associated with the spa or hot tub water circulating system, including pump motors.

3. Metal conduit and metal piping that is located within 5 feet of the inside walls of the spa or hot tub and that are not separated from the spa or hot tub by a permanent barrier.

(4) All metal surfaces that are located within 5 feet of the inside walls of a spa or hot tub and not separated from the spa or hot tub area by a permanent barrier.

(5) Electrical devices and controls not associated with the spas or hot tubs must be located at least 5 feet away from such units or are required to be bonded to the spa or hot tub system.

GFCI protection of spa or hot tub installed indoors or outdoors.

Section 680-42 contains GFCI requirements that apply to both indoor and outdoor spa and hot tub equipment. The general requirement is that, "An outlet that supplies a self-contained spa or hot tub, or a packaged spa or hot tub equipment assembly, shall be protected by a ground-fault circuit-interrupter."*

An exception permits the GFCI protection at the outlet to be eliminated for a listed self-contained unit or listed pack-

GFCI Protection of Spas and Hot Tubs

Panelboard

GFCI protection of outlet required

Outlet

Exception: Listed self-contained unit or listed packaged equipment assembly that is marked to indicate integral GFCI protection is provided for all electrical parts within the unit or assembly.

Figure 16-25

aged equipment assembly that is marked to indicate that integral ground-fault circuit-interrupter protection is provided for all electrical parts within the unit or assembly (pumps, air blowers, heaters, lights, controls, sanitizer generators, wiring, etc.).

Electric signs and outline lighting.

Section 600-7 contains requirements for grounding of electric signs and metal equipment of outline lighting systems. The rules in Article 600 amend or supplement the general rules for grounding in

Grounding Signs and Outline Lighting

- Signs and metal equipment of outline lighting shall be grounded.
- Listed flexible metal conduit to be terminated in connector that ensures positive bonding.
- Bonding conductor for flexible nonmetallic conduit or tubing to be external, not smaller than No. 12 copper per Section 250-95.

Figure 16-26

Article 250 due to the rule of "Code Arrangement" in Section 90-3.

Specific requirements in Section 600-7 are that, "Signs and metal equipment of outline lighting systems shall be grounded." Basically, there are no exceptions to this requirement, so even small parts like metal-mounting feet of glass-tube supports are required to be grounded.

Section 600-7 continues, "Listed flexible metal conduit that encloses the secondary wiring of a transformer or power supply for use with electrical discharge tubing shall be permitted as a grounding means provided the flexible metal conduit terminates in a connector that ensures positive bonding connection." No maximum length of listed flexible metal conduit is given in this section that may be used for the grounding conductor on the secondary of a sign transformer or power supply. Some interpret this to mean that, since Section 600-7 does not state a maximum length, the six-foot maximum length given in Section 250-91(b) Exception No. 1 and in Section 350-14 Exception applies. Others maintain that since no maximum length is given in Section 600-7, listed flexible metal conduit can be used for grounding on the secondary of sign transformers and power supplies without length restriction.

"Where flexible nonmetallic conduit or tubing is used to enclose the secondary wiring of a transformer or power supply and a bonding conductor is required to bond metal electrode receptacles and other metal parts of a sign, the bonding conductor shall be:

(1) Installed on the exterior of the flexible nonmetallic tubing or conduit; and

(2) Copper not smaller than No. 12 in accordance with Section 250-95."

Chapter Sixteen: The questions included here were developed using material included in this chapter. The answers can be found by reviewing the text. It is also important that students make use of the 1996 NEC®, where many answers can be found. See pages 237 - 238 for answers.

1. The Code requires that metal raceways, enclosures, frames and other noncurrent-carrying metal parts be kept at least ____ away from lightning rod conductors.
   a. 4 feet
   b. 6 feet
   c. 8 feet
   d. 12 feet

2. Where it is not practicable to obtain the required separation between the metal enclosure and the lightning rod conductor, they are required to be ____.
   a. grounded to building steel
   b. protected by an impedance device
   c. bonded together
   d. connected to a ground rod

3. Generally, in hazardous (classified) locations, locknuts installed on each side of an enclosure, or a locknut on the outside and a bushing on the inside ____ be used for bonding.
   a. shall
   b. can
   c. cannot
   d. may

4. Bonding locknuts or bonding bushings with bonding jumpers must be used in hazardous (classified) locations to ensure the integrity of the bond and its capability of carrying any ____ that may be imposed.
   a. unidentified currents
   b. neutral currents
   c. objectionable current
   d. fault current

5. Where flexible metal conduit is permitted and used in a hazardous (classified) location, it must have an internal or external bonding jumper installed to supplement the conduit. Where the jumper is installed on the outside, it is limited to ____ feet in length.
   a. 6
   b. 7
   c. 8
   d. 9

6. Where equipment is installed in an agricultural building that requires grounding, a ____ equipment grounding conductor must be installed to ground the equipment.
   a. aluminum
   b. copper
   c. copper-clad
   d. steel

7. Minimizing or elimination of stray voltages in an agricultural building is accomplished by bonding together, with a copper conductor not smaller than No. ____, the reinforcing rods and wire mesh in the concrete and then connecting these to the metal piping systems, stanchions and grounding electrode systems.
   a. 8
   b. 6
   c. 4
   d. 2

8. A listed impedance device that essentially consists of a saturable core reactor is called a ____ and is permitted to be used to bond interior metal piping systems in agricultural buildings.
   a. bonding strap
   b. grounding wedge
   c. grounding device
   d. bonding isolator

9. A redundant means of grounding is required in ____ of health care facilities.
   a. limited-care facilities
   b. office areas
   c. all areas
   d. patient care areas

10. Where used in patient care areas, Type MI cable or Type ____ cable is required to have outer metal armor or sheath that is ____ as an acceptable grounding return path.
   a. MV - approved by special permission
   b. SE - permitted
   c. SNM - approved
   d. MC - identified

11. Where metal raceways or Type MC or MI cables are used for feeders supplying critical care areas in health care facilities, grounding of panelboards and switchboards must be assured. Acceptable methods include all of the following EXCEPT ____
   a. grounding bushings with bonding jumpers
   b. threaded bosses
   c. bonding locknuts or bushings
   d. standard locknuts

12. Equipment in the vicinity of swimming pools must be bonded together with a No. 8 or larger copper conductor. This also includes metal raceways and cables, piping and fixed metal parts within ____ feet horizontally or ____ feet vertically of the pool, observation stands, towers, platforms or diving structures.
   a. 5 - 18
   b. 6 - 20
   c. 5 - 12
   d. 9 - 14

13. A common bonding grid at a pool is permitted to be the structural reinforcing steel of a concrete pool, the wall of a bolted or welded pool, or a No. ____ or larger solid copper conductor.
   a. 8
   b. 6
   c. 4
   d. 2

14. A dry-niche lighting fixture is required to be provided with a provision for the drainage of water and a means for accommodating ____ equipment grounding conductor(s) for each conduit entry.
   a. 1
   b. 2
   c. 3
   d. 0 or none

15. All 125-volt receptacles located within ____ feet of the inside walls of a pool are required to be protected by a ground-fault circuit-interrupter.
   a. 30
   b. 25
   c. 22
   d. 20

16. Ground-fault circuit-interrupters used around pools are permitted to be installed in a ____ that contains circuits protected by other than ground-fault circuit-interrupters.
   a. cabinet
   b. metal device box
   c. panelboard
   d. nonmetallic box

17. For other than storable pools, a flexible cord is not permitted to exceed ____ feet in length and must have a copper equipment grounding conductor not smaller than No. 12 with a grounding-type attachment plug.
   a. 3
   b. 4
   c. 5
   d. 6

18. Ground-fault circuit-interrupters used around pools are required to be self-contained units, circuit-breaker types, receptacle-types or other ____ types.
   a. listed
   b. acceptable
   c. approved
   d. labeled

19. A ground-fault circuit-interrupter is required to be installed in the branch circuit supplying underwater lighting fixtures operating at more than ____ volts, so that there is no shock hazard during relamping.
   a. 6
   b. 10
   c. 12
   d. 15

20. The installation of the ground-fault circuit-interrupter installed to protect an underwater lighting fixture must be such that there is no shock hazard with any likely fault-condition combination that involves a person in a conductive path from any ungrounded part of the ____ or the fixture to ground.
   a. service
   b. system
   c. feeder
   d. branch circuit

21. Where metal conduit is used to supply a ____ niche fixture, it is required to be of brass or other approved corrosion-resistant metal.
   a. dry-
   b. wet-
   c. no-
   d. metal-

22. A wet-niche fixture is required to be bonded to and secured to the forming shell by a positive locking device that assures a low-resistance contact and requires a ____ to remove the fixture from the forming shell.
   a. tool
   b. permit
   c. fitting
   d. helper

23. Transformers used for the supply of underwater fixtures, together with the transformer enclosure, are required to be ____ for the purpose.
   a. suitable
   b. identified
   c. acceptable
   d. approved

24. The transformer used to supply underwater lighting fixtures is required to be a ____ winding type having a grounded metal barrier between the primary and secondary windings.
   a. three
   b. isolated
   c. one
   d. four

25. Signs and ___ must be grounded.
   a. all line-side equipment
   b. electrode receptacle housings
   c. metal equipment of outline lighting
   d. electrical-discharge tubing

26. Where flexible nonmetallic conduit is used on the secondary of a sign transformer, the bonding conductor must be:
   a. installed inside the conduit
   b. not more than 6-feet long
   c. installed outside the conduit
   d. listed GTO cable

Notes:

17

Over 600 Volt Systems

Objectives

After studying this chapter, the student will be able to understand

- Grounding rules for high voltage (1 kV and over).
- Use of surge arresters.
- Grounding of outdoor industrial substations.

High voltage (1 kV and over).

For grounding of systems or circuits of 1kV and over, the NEC® indicates that where such high-voltage systems are grounded, they shall comply with Article 250 plus other rules in Part M of Article 250 which supplement or modify other rules in Article 250.

Most medium-voltage systems (2.4 to 15kV) are resistance grounded while others are high resistance grounded. The only difference in the various methods of resistance grounding is the amount of ground-fault current that will flow due to the varying resistance.

Most high-voltage systems (above 15kV) are effectively grounded utilizing surge arresters because the transmission lines are open and commonly subjected to lightning surges and transient overvoltages.

Systems rated 2,400 volts to 13,800 volts.

Grounding may be solid or low resistance depending on the available ground-fault current and the size of the system. Reactance grounding in this voltage range is preferred if the circuits are overhead and thus subject to lightning exposure, if they serve rotating machinery and if excessive ground-fault current will not develop. It is common in industrial systems to ground the neutral of systems rated 2,400 volts and above through a resistor.

Owing to the higher voltage of these systems, compared to systems of 600 volts and less, the ground-fault current levels that will flow in the grounding electrode conductor and grounding electrode is sufficiently high that a current transformer may be placed in the grounding electrode conductor to give positive indication that a ground fault exists or may be used to clear the circuit in the event the ground fault persists over a predetermined time. Compare this with a system of 600 volts or less. There, the current transformer cannot be placed satisfactorily in the grounding electrode conductor, for little ground-fault current will flow in that conductor.

Systems rated 15,000 volts or more.

Such systems are nearly always outdoors and have no rotating equipment served by them. Usually they are solidly grounded. Such solid grounding permits the use of grounded-neutral type surge arresters, which cost less as well as provide better protection from overvoltage.

In systems over 600 volts, there will be different voltage levels for various utilization purposes. Grounding must be resorted to at each voltage level as previously described for systems of 600 volts or less.

Grounding outdoor industrial substations.

The grounding of outdoor industrial substations is concerned with grounding the fence and other supporting structures, all equipment and conductor enclosures and the grounding of the neutral.

A ground bus or grid should be established extending about 3 feet outside the periphery of the fence. The bus should be connected to many ground rods around its periphery and, in addition, should be connected to a metallic underground water piping system where available. The ground bus should have a minimum size of No. 4/0 and be approximately 25 percent of the capacity of the system. Cable may be sized on the basis of the capacity of bare conductor in free air and copper bus on the basis of 1,200 amperes per square inch. Connections from the fence and from all equipment within the fence, that is, transformer cases, steel structures, switchgear including operating mechanisms for gang-operated disconnects, etc., should be not less than No. 4/0 nor less than 25 percent of the capacity of the secondary conductors. To be sure of having a good, permanent neutral ground, it is best to connect the neutral to two points on the ground bus.

A grounding bus and connections are effective only if the mechanical construction is sound and as permanent as possible. No connection should be soldered. It is preferable to properly braze or weld all connections and to protect all cable from mechanical injury. If a metallic enclosure is used for the mechanical protection of the conductor, the enclosure should be connected in parallel with the cable being protected from end to end of the enclosure.

There is some difference of opinion as to whether the fence ground should be separated from the ground bus used for the station. If the fence is connected to the station bus and a fault occurs, the fence will be elevated above earth potential by the IZ drop. The potential gradient in the earth near the fence drops off very rapidly within the first few feet. Anyone contacting the fence may thus be subjected to a hazardous step voltage under fault conditions.

If the fence ground was isolated from the station ground, then the fence will be elevated from the station ground, and under fault conditions anyone contacting the fence and the equipment will be subjected to a hazard.

In most industrial substations, the fence and equipment are close. In such cases, it is best to connect the fence ground and the station ground bus. The shock hazard when using a common ground for fence and equipment can be kept to a minimum by having the ground connection resistance as low as possible.

Section 250-151 permits a system neutral to be derived from an ungrounded system by a grounding transformer (zigzag, wye-delta, or T-connected). Ground-fault protection may be provided, and positive tripping can be accomplished with low magnitudes of ground-fault current.

Section 250-152 also permits solidly grounded neutral systems. Multiple grounding is permitted at more than one point for services, direct buried portions of feeders employing a bare copper neutral and for overhead portions of a system.

Section 250-153 permits impedance grounded neutral systems. Such impedance grounding can be accomplished by reactance grounding, resistance grounding or high-resistance grounding.

Section 250-155 generally requires grounding of all noncurrent-carrying metal parts of fixed, portable or mobile equipment of high voltage equipment including fences, housings, enclosures and supporting structures.

Exception No. 1 excludes from the grounding requirement equipment that is isolated from ground and located so as to prevent any person who can make contact with the ground from contacting such metal parts when the equipment is energized.

Exception No. 2 excludes from the grounding requirement pole-mounted distribution apparatus as provided in Serction 250-42, Exception No. 3. There, the equipment is required to be at a height exceeding 8 feet above ground or grade level.

Surge arresters.

Where surge arresters are used on secondary services of less than 1,000 volts, it is required that the connections to the service conductor and to the grounding conductor be as short as practicable. The Code specifies that the grounding conductor may be (1) the grounded service conductor; (2) the common grounding electrode conductor; (3) the service grounding electrode or (4) the equipment grounding terminal in the service equipment. The use of method (3) is practical only on an ungrounded system. A minimum size of No. 14 copper or No. 12 aluminum is specified. Note that only the minimum size is specified.

If a surge arrester protects the primary of a transformer which supplies a secondary distribution system, the grounding conductor may be interconnected to the secondary neutral, provided that in addition to the direct grounding connection at the arrester, the grounded conductor of the secondary has elsewhere a grounding connection to a continuous metallic underground water piping system. However, where there are not less than four secondary connections to any underground metallic water piping system per mile, the direct grounding connection at the arrester may be omitted.

The grounding electrode conductor of a lightning arrester also may be connected to the secondary neutral. This is true if, in addition to the direct grounding connection at the arrester, the grounding conductor of the secondary system is part of a multi-grounded neutral system of which the primary neutral has at least four grounding connections in each mile of line, as well as a ground at each service.

Where the secondary is not grounded to a metallic water system but uses other available electrodes, an interconnection between surge arrester grounding conductor and the secondary neutral shall be made through a spark gap. The spark gap must have a breakdown voltage of at least twice the primary circuit voltage but not necessarily more than 10 kV. There shall be at least one other ground on the grounded conductor of the secondary not less than 20 feet away from the surge arrester grounding electrode.

No other connection between a surge arrester ground and a secondary neutral is allowed by the Code except by special permission.

Chapter Seventeen: The questions included here were developed using material included in this chapter. The answers can be found by reviewing the text. It is also important that students make use of the 1996 NEC®, where many answers can be found. See page 238 for answers.

1. Most medium-voltage systems ____ to ____ are resistance grounded while others are high resistance grounded.
   a. 5.4 - 20kV
   b. 2.4 - 15kV
   c. 3.3 - 17kV
   d. 4.6 - 18kV

2. Most high-voltage systems ____ are effectively grounded utilizing surge arresters because the transmission lines are open and commonly subjected to lightning surges and transient overvoltages.
   a. above 15kV
   b. above 12kV
   c. above 10kV
   d. above 11kV

3. It is common in industrial systems to ground the neutral of systems rated ____ volts and above through a resistor.
   a. 2,000
   b. 2,400
   c. 1,800
   d. 1,380

4. Systems rated ____ volts or more are nearly always outdoors and have no rotating equipment served by them.
   a. 10,000
   b. 12,000
   c. 15,000
   d. 13,000

5. In systems over ____ volts, there will be different voltage levels for various utilization purposes.
   a. 1,000
   b. 2,000
   c. 600
   d. 1,500

6. The grounding of outdoor industrial substations is concerned with grounding the fence and other supporting structures, all equipment and conductor enclosures and the grounding of the neutral. A ground bus or grid should be established extending about ____ feet outside the periphery of the fence.
   a. 3
   b. 4
   c. 6
   d. 8

7. Cable can be sized on the basis of the capacity of bare conductor in free air and copper bus on the basis of ____ amperes per square inch.
   a. 1,250
   b. 1,000
   c. 1,200
   d. 1,300

8. Connections from the fence and from all equipment within the fence, that is, transformer cases, steel structures, switchgear including operating mechanisms for gang-operated disconnects, etc., should be not less than No. ____ nor less than ____ percent of the capacity of the secondary conductors.
   a. 3/0 - 50
   b. 4/0 - 25
   c. 2/0 - 40
   d. 1/0 - 30

9. To be sure of having a good, permanent neutral ground, it is best to connect the neutral to ____ points on the ground bus.
   a. four
   b. three
   c. two
   d. five

10. Impedance grounded neutral systems are permitted and can be accomplished by all but one of the following methods:
   a. reactance grounding
   b. resistance grounding
   c. high-resistance grounding
   d. low-resistance grounding

11. Without exception, where surge arresters are installed either indoors or outdoors they are required to be ____ to unqualified persons.
   a. inaccessible
   b. accessible
   c. readily accessible
   d. available

12. Where there are not less than four secondary connections to any underground metallic water piping system per each ____, the direct grounding connection at the arrester may be omitted.
   a. ¼ mile
   b. mile
   c. ½ mile
   d. ¾ mile

Chapter 18

Tables

Objectives

After studying this chapter, the reader will be able to understand

- Application of tables to determine proper types and sizing of grounding electrode conductor.
- How to determine where voltage drop will affect sizes.
- Determine minimum size of equipment grounding conductors.
- Dimensions and lengths of conduits and other raceways.

Included in this chapter are several tables that are vital or useful in studying the subject of grounding and correctly applying the NEC® rules in that regard.

In addition, several tables are included to assist the installer or inspector of electrical systems in properly designing, installing and maintaining these systems.

By carefully utilizing the information in these tables, the electrical system will benefit from additional safety and should serve the owner for years to come.

The following tables are included:

Table 1. Table 250-94, Grounding Electrode Conductor for AC systems from the NEC®.

Table 2. Analysis of Table 250-94 grounding electrode conductor and service conductor compared as to relative size.

Table 3. Comparison of copper grounding electrode conductor per Table 250-94 when used with and without steel conduit for physical protection.

Table 4. Comparison of aluminum grounding electrode conductors per Table 250-94 when used with and without aluminum conduit for protection.

Table 5. Rating of grounding electrode conductors specified in Table 250-94 and voltage drop under maximum short-time rating.

Table 6. Table 250-95 from the National Electrical Code® "Minimum size Equipment Grounding Conductors for Grounding Raceways and Equipment."

Table 7. Analysis of Table 250-95 of the National Electrical Code®.

Table 8. Table 8 from Chapter Nine of the National Electrical Code® "Conductor Properties."

Table 9. Not used.

Table 10. Not used.

Table 11. Maximum length of electrical metallic tubing, intermediate metal conduit and rigid steel conduit that may safely be used as an equipment grounding circuit conductor.

Table 12. Not used.

Table 13. Not used.

Table 14. Aluminum conduit used as an equipment grounding conductor. Compared with equipment grounding conductors as specified in Table 250-95 of the National Electrical Code®.

Table 15. Sizes of conductor (copper bar) for use in the equipment grounding strap for overcurrent devices from 1,600 amperes through 5,000 amperes.

Table One

Table 250-94
Grounding Electrode Conductor for AC Systems

Size of Largest Service-Entrance Conductor or Equivalent Area for Parallel Conductors		Size of Grounding Electrode Conductor	
Copper	Aluminum or Copper-Clad Aluminum	Copper	*Aluminum or Copper-Clad Aluminum
2 or smaller	1/0 or smaller	8	6
1 or 1/0	2/0 or 3/0	6	4
2/0 or 3/0	4/0 or 250 kcmil	4	2
Over 3/0 thru 350 kcmil	Over 250 kcmil thru 500 kcmil	2	1/0
Over 350 kcmil thru 600 kcmil	Over 500 kcmil thru 900 kcmil	1/0	3/0
Over 600 kcmil thru 1100 kcmil	Over 900 kcmil thru 1750 kcmil	2/0	4/0
Over 1100 kcmil	Over 1750 kcmil	3/0	250 kcmil

Where multiple sets of service-entrance conductors are used as permitted in Section 230-40, Exception No. 2, the equivalent size of the largest service-entrance conductor shall be determined by the largest sum of the areas of the corresponding conductors of each set.

Where there are no service-entrance conductors, the grounding electrode conductor size shall be determined by the equivalent size of the largest service-entrance conductor required for the load to be served.

*See installation restrictions in Section 250-92(a).

(FPN): See Section 250-23(b) for size of alternating-current system grounded conductor brought to service equipment.

Table Two

Analysis of Table 250-94. *Grounding Electrode Conductor and Service Conductor Compared As To Relative Size				
Size of Copper Service Conductor AWG No.	Continuous Rating of Service Conductor 75° C Wire Ampacity	Size of Copper Grounding Electrode Conductor AWG No.	Continuous Rating Grounding Electrode Conductor 75° C Wire Ampacity	Grounding Electrode Conductor Expressed as a percent of Service Conductor
2 or smaller	115	8	50	43%
0	150	6	65	43%
3/0	200	4	85	43%
350 kcmil	310	2	115	37%
600 kcmil	420	1/0	150	36%
1000 kcmil	545	2/0	175	32%
1250 kcmil (over 1100 kcmil)	590	3/0	200	34%

* This grounding Electrode Conductor is that part of the Grounding System which is the sole connection between the grounding electrode and the grounded system conductor.

Table Three

Comparison of Copper Grounding Electrode Conductors per Table 250-94 When Used With and Without Steel Conduit for Physical Protection					
Copper Wire Size. AWG No.	DC Resistance of Bare Wire. Ohms/100'	Size of conduit to enclose conductor. Inch	*Approximate DC Resistance of conduit. Ohms/100'	*Approximate Impedance of Conductor in Conduit. Ohms/100' at 500 Amps Sq. In. density	
				Conduit bonded at both ends.	Conduit NOT bonded.
8	0.07780	½	0.0321	0.105	0.210
6	0.04910	¾	0.0242	0.078	0.156
4	0.03080	¾	0.0242	0.078	0.156
2	0.01940	1	0.0154	0.056	0.112
0	0.01220	1	0.0154	0.056	0.112
00	0.00967	1	0.0154	0.056	0.112
000	0.00766	1¼	0.0120	0.044	0.088

* These values are approximate but sufficiently accurate for all practical purposes. Manufacturing tolerances for conduit, although quite satisfactory for its physical requirements, can vary the resistance up to about 15 percent compared to standard dimensions.

Table Four

Comparison of Aluminum Grounding Electrode Conductors per Table 250-94 When Used With and Without Aluminum Conduit for Physical Protection			
Aluminum Wire Size	DC Resistance Ohms/100'	*Size of Conduit to Enclose Conductor, In.	Approximate DC Resistance of Aluminum Conduit Ohms/100'
6	0.08080	½	0.00800
4	0.05080	½	0.00800
2	0.03190	¾	0.00605
1/0	0.02010	1	0.00385
3/0	0.01260	1	0.00385
4/0	0.00100	1¼	0.00300
250 kcmil	0.00847	1¼	0.00300

*Must be installed to comply with 250-92(a), that is, made electrically continuous from the point of attachment to cabinets or equipment to the grounding electrode and shall be securely fastened to the ground clamp.

Table Five

Rating of Grounding Electrode Conductors Specified in Table 250-94 and Voltage Drop under Maximum Short-Time Rating

Grounding Conductor Size AWG	DC Resistance per 100 ft. Copper Conductor	Ampacity 75° C Wire Continuous Rating	Circular Mils	*Short-Time Rating Amps	Ratio Short-Time Rating to Continuous Rating	Voltage Drop per 100 ft. at Short-Time Rating
8	0.077800	50	16510	391	7.8	30.4
6	0.049100	65	26240	621	9.6	30.5
4	0.030800	85	41740	988	11.6	30.4
3**	0.024500	100	52620	1,245	12.5	30.5
2	0.019400	115	66360	1,571	13.7	30.5
1**	0.015400	130	83690	1,981	15.2	30.5
1/0	0.012200	150	105600	2,499	16.7	30.5
2/0	0.009670	175	133100	3,150	18.0	30.5
3/0	0.007660	200	167800	3,972	19.9	30.4
4/0**	0.006080	230	211600	5,008	21.8	30.5
kcmil						
250**	0.005150	255	250000	5,917	23.2	30.5
300**	0.004290	285	300000	7,101	24.9	30.5
350**	0.003670	310	350000	8,284	26.7	30.4
400**	0.003210	335	400000	9,467	28.3	30.4
500**	0.002580	380	500000	11,834	31.1	30.5

* Based on 1 ampere per 42.25 of circular mil area for five seconds.

** Not shown in Table 250-94 but listed here to assist in sizing grounding conductor lengths over 100 feet long.

If length of grounding conductor exceeds 100 feet, the conductor should be increased in size so that voltage drop based on short-time current rating of the conductor specified should not exceed 40 volts. In other words, the resistance of the grounding conductor used when over 100 feet long should not exceed the resistance of 100 feet of the conductor specified in the table.

Table Six

Table 250-95. Minimum Size Equipment Grounding Conductors for Grounding Raceway and Equipment		
Rating or Setting of Automatic Overcurrent Device in Circuit Ahead of Equipment, Conduit, etc., Not Exceeding (Amperes)	Size	
	Copper Wire No.	Aluminum or Copper-Clad Aluminum Wire No.*
15	14	12
20	12	10
30	10	8
40	10	8
60	10	8
100	8	6
200	6	4
300	4	2
400	3	1
500	2	1/0
600	1	2/0
800	1/0	3/0
1000	2/0	4/0
1200	3/0	250 kcmil
1600	4/0	350 kcmil
2000	250 kcmil	400 kcmil
2500	350 kcmil	600 kcmil
3000	400 kcmil	600 kcmil
4000	500 kcmil	800 kcmil
5000	700 kcmil	1200 kcmil
6000	800 kcmil	1200 kcmil

*See installation restrictions in Section 250-92(a).

Note: Equipment grounding conductors may need to be sized larger than specified in this table in order to comply with Section 250-51.

Table Seven

Analysis of Table 250-95

Rating of OC Device in Amperes	EGC Copper Conductor		*Short-Time Rating of EGC Conductor in Amps	** K-factor	Percent Rating of EGC Conductor to OC Device	*** Ampacity per Table 310-16
	AWG No.	Cir. Mils.				
20	12	6530	155	7.7	125%	25
30	10	10380	246	8.2	117%	35
40	10	10380	246	6.1	88%	35
60	10	10380	246	4.1	58%	35
100	8	16510	391	3.9	50%	50
200	6	26240	621	3.1	33%	65
400	3	52620	1,245	3.1	25%	100
600	1	83690	1,981	3.3	22%	130
800	1/0	105600	2,499	3.1	19%	150
1000	2/0	133100	3,150	3.2	18%	175
1200	3/0	167800	3,972	3.3	17%	200
1600	4/0	211600	5,008	3.1	14%	230
2000	250 kcmil	250000	5,914	3.0	13%	255
2500	350 kcmil	350000	8,285	3.3	12%	310
3000	400 kcmil	400000	9,467	3.2	11%	335
4000	500 kcmil	500000	11,834	3.0	10%	380
5000	700 kcmil	700000	16,568	3.3	9%	460
6000	800 kcmil	800000	18,935	3.2	8%	490

*   One ampere per 42.25 circular mils for five seconds.

**   K factor. Ampere rating of overcurrent device times K equals short-time rating of E.G.C. conductor in amperes.

***   Based on 75° C copper wire, Table 310-16.

Table Eight

		Conductors				DC Resistance at 75°C (167°F)		
		Stranding		Overall		Copper		Aluminum
Size AWG/ kcmil	Area Cir. Mils	Quan-tity	Diam. In.	Diam. In.	Area In.²	Uncoated ohm/kFT	Coated ohm/kFT	ohm/ kFT
18	1620	1	-	0.040	0.001	7.77	8.08	12.8
18	1620	7	0.015	0.046	0.002	7.95	8.45	13.1
16	2580	1	-	0.051	0.002	4.89	5.08	8.05
16	2580	7	0.019	0.058	0.003	4.99	5.29	8.21
14	4110	1	-	0.064	0.003	3.07	3.19	5.06
14	4110	7	0.024	0.073	0.004	3.14	3.26	5.17
12	6530	1	-	0.081	0.005	1.93	2.01	3.18
12	6530	7	0.030	0.092	0.006	1.98	2.05	3.25
10	10380	1	-	0.102	0.008	1.21	1.26	2.00
10	10380	7	0.038	0.116	0.011	1.24	1.29	2.04
8	16510	1	-	0.128	0.013	0.764	0.786	1.26
8	16510	7	0.049	0.146	0.017	0.778	0.809	1.28
6	26240	7	0.061	0.184	0.027	0.491	0.510	0.808
4	41740	7	0.077	0.232	0.042	0.308	0.321	0.508
3	52620	7	0.087	0.260	0.053	0.245	0.254	0.403
2	66360	7	0.097	0.292	0.067	0.194	0.201	0.319
1	83690	19	0.066	0.332	0.087	0.154	0.160	0.253
1/0	105600	19	0.074	0.372	0.109	0.122	0.127	0.201
2/0	133100	19	0.084	0.418	0.137	0.0967	0.101	0.159
3/0	167800	19	0.094	0.470	0.173	0.0766	0.0797	0.126
4/0	211600	19	0.106	0.528	0.219	0.0608	0.0626	0.100
250	-	37	0.082	0.575	0.260	0.0515	0.0535	0.0847
300	-	37	0.090	0.630	0.312	0.0429	0.0446	0.0707
350	-	37	0.097	0.681	0.364	0.0367	0.0382	0.0605
400	-	37	0.104	0.728	0.416	0.0321	0.0331	0.0529
500	-	37	0.116	0.813	0.519	0.0258	0.0265	0.0424
600	-	61	0.099	0.893	0.626	0.0214	0.0223	0.0353
700	-	61	0.107	0.964	0.730	0.0184	0.0189	0.0303
750	-	61	0.111	0.998	0.782	0.0171	0.0176	0.0282
800	-	61	0.114	1.030	0.834	0.0161	0.0166	0.0265
900	-	61	0.122	1.094	0.940	0.0143	0.0147	0.0235
1000	-	61	0.128	1.152	1.042	0.0129	0.0132	0.0212
1250	-	91	0.117	1.289	1.305	0.0103	0.0106	0.0169
1500	-	91	0.128	1.412	1.566	0.00858	0.00883	0.0141
1750	-	127	0.117	1.526	1.829	0.00735	0.00756	0.0121
2000	-	127	0.126	1.632	2.092	0.00643	0.00662	0.0106

These resistance values are valid ONLY for the parameters as given. Using conductors having coated strands, different stranding type, and especially, other temperatures, changes the resistance.

Formula for temperature change: $R_2 = R_1 [1+a(T_2-75)]$ where: $a_{cu} = 0.00323, a_{AL} = 0.00330$.

Conductors with compact and compressed stranding have about 9 percent and 3 percent, respectively, smaller bare conductor diameters than those shown. See Table 5A for actual compact cable dimensions.

The IACS conductivities used: bare copper = 100%, aluminum = 61%.

Class B stranding is listed as well as solid for some sizes. Its overall diameter and area is that of its circumscribing circle. (FPN): The construction information is per NEMA WC8-1988. The resistance is calculated per National Bureau of Standards Handbook 100, dated 1966, and Handbook 109, dated 1972.

Table Eleven

Maximum length of steel conduit or tubing that may safely be used as an equipment grounding circuit conductor						
Conduit Size In.	AWG/ kcmil	OC 75°C	Fault 500%	EMT Length	IMC Length	RGC Length
½	14	15	75	231	232	227
	12	20	100	246	255	247
	10	30	150	226	250	240
¾	12	20	100	266	262	257
	10	30	150	255	259	250
	8	40	200	261	275	264
	8	50	250	211	229	222
1	8	40	200	288	284	275
	8	50	250	236	237	229
	6	60	300	263	270	257
	6	70	350	228	240	228
	4	80	400	254	274	259
	4	90	450	228	251	238
	3	100	500	227	256	244
1¼	3	100	500	268	274	258
	2	110	550	273	283	265
	2	125	625	244	257	243
	1	125	625	266	281	263
	1	150	750	226	245	234
1½	1	125	625	285	292	273
	1	150	750	243	256	241
	1/0	150	750	264	279	260
2	2/0	175	875	280	287	268
	3/0	200	1000	268	281	261
	4/0	225	1125	257	277	257
	4/0	250	1250	234	257	240
2½	250	250	1250	289	269	261
	250	275	1375	268	252	244
	300	275	1375	280	263	254
	300	300	1500	261	248	239

Based on a clearing ground-fault current of 500 percent of overcurrent device rating; circuit 120 volts to ground; 50 volts drop at the arc; 30° C ambient temperature; 75° C conductor temperature.
Calculated with "Steel Conduit Analysis Vs 1.2" by Georgia Institute of Technology.

Table Eleven (continued)

Conduit Size In.	AWG/ kcmil	OC 75°C	Fault 500%	EMT Length	IMC Length	RGC Length
2½	400	300	1500	278	263	253
	400	350	1750	244	239	227
	500	350	1750	255	249	236
	500	400	2000	227	230	216
3	500	350	1750	280	262	253
	500	400	2000	251	240	231
	600	400	2000	260	248	238
	600	450	2250	235	231	220
	700	450	2250	242	237	225
	700	500	2500	221	222	209
3½	600	400	2000	280	256	248
	600	450	2250	257	237	229
	700	450	2250	264	244	235
	700	500	2500	243	228	218
	750	450	2250	267	246	237
	750	500	2500	246	230	221
	900	500	2500	254	237	227
	900	600	3000	220	212	200
4	750	450	2250	277	253	245
	750	500	2500	255	236	228
	900	500	2500	264	243	234
	900	600	3000	229	216	206
	1000	500	2500	269	247	238
	1000	600	3000	233	219	210

Based on a clearing ground-fault current of 500 percent of overcurrent device rating; circuit 120 volts to ground; 50 volts drop at the arc; 30° C ambient temperature; 75° C conductor temperature.
Calculated with "Steel Conduit Analysis Vs 1.2" by Georgia Institute of Technology.

Table Fourteen

**Aluminum Conduit
used as an equipment grounding circuit conductor (EGC)
compared with equipment grounding circuit (EGC) conductors
as specified in Table 250-95 of the National Electrical Code®**

Size of Conduit In.	Area of Conduit Wall Sq. In.	Area of Conduit Wall Based on NEMA Standards Sq. In.	DC Resistance Ohms/M ft.	Largest EGC Conductor as specified in Table 250-95 Aluminum Wire No.	Nearest Wire No. Equivalent to Conduit	
					Aluminum Wire No.	Copper Wire No.
½	0.254	0.254	0.08000	12	4/0	2/0
¾	0.337	0.337	0.06050	8	300 kcmil	4/0
1	*0.530	0.500	0.03850	6	500 kcmil	300 kcmil
1¼	*0.680	0.670	0.03000	4	600 kcmil	350 kcmil
1½	*0.790	0.790	0.02580	4	700 kcmil	500 kcmil
2	*1.030	1.080	0.01980	1	900 kcmil	600 kcmil
2½	1.710	1.710	0.01190	1	1500 kcmil	1000 kcmil
3	*2.310	2.250	0.00880	2/0	2000 kcmil	1250 kcmil
3½	2.700	2.700	0.00752	2/0	over	1500 kcmil
4	*3.400	3.180	0.00598	2/0	2000 kcmil	2000 kcmil
5	4.300	4.300	0.00473	3/0	over 2000 kcmil	over 2000 kcmil

*Values as measured from a sample.

Column 3 gives values as calculated on basis of NEMA standards.

Table Fifteen

Sizes of conductor (Copper Bar) for use in the equipment grounding circuit and for the Equipment Grounding Strap for overcurrent devices from 1600 amperes through 5000 amperes						
Rating of OC Device in Amps.	Copper EG Conductor		* Conductor Rating Amps	** Short-Time Rating of EGC Conductor in Amps	*** K factor	**** Percent Rating of EGC Conductor to OC Device
	Copper Bar - Inch	Cir. Mils.				
1600	¼ x 1	318300	366	10,610	5.3 (6.6)	22.9
2000	¼ x 1¼	397900	457	13,260	6.0 (6.6)	22.9
2500	¼ x 1½	477450	549	15,920	6.0 (6.4)	21.9
3000	¼ x 2	636600	647	21,220	6.0 (7.1)	21.6
4000	¼ x 2½	795750	809	26,525	6.3 (6.6)	20.2
5000	¼ x 4	1273000	1,220	42,430	6.8 (8.5)	24.4

* Based on 30° C rise, 40° C ambient.

** One ampere per 30 circular mils for 5 seconds. Use this value ONLY where insulated, current-carrying conductors will NOT come in contact with the bare equipment grounding bus bars. Where there is likelihood of contact, then recalculate the 5 second withstand rating on the basis of one ampere...for five seconds...for every 42.25 circular mils of copper conductor. This is discussed in Chapter 11 of this text.

*** K-factor. Ampere rating of O. C. device times K-factor equals short-time rating of EGC conductor in amperes. Figures in brackets are based on the actual EGC conductor used.

**** Based on 30° C rise, 40° C ambient for EGC conductor.

Appendix

The History
and Mystery
of Grounding

In 1941, Howard S. Warren completed work on the manuscript "How The Code Came." This material was turned over to then president, Alvah Small, Underwriters Laboratories, to be made available to anyone should it serve a useful purpose. Alvah Small was chairman of the NFPA Electrical Committee from 1925 to 1950. This tome on the history of the Code was dedicated to Alvah for his work as chairman in keeping the Code abreast of developments in the electrical industry and always in the public interest. Warren represented the telephone group on the Electrical Committee from 1907 until 1939. Warren was preceded by C.J.H. Woodbury and followed by A.H. Schirmer, L.S. Inskip and L.H. Sessler.

The subject of grounding has always been confusing to many, and as such, precipitates much interest wherever sessions are held on Code discussions. Grounding has always had its controversial aspects. Part of this stems from the fact that the early sponsors of the Code, the Underwriters' National Electric Association, represented the insurance interests and the grounding of electrical circuits had much to offer for personnel safety, but little fire safety for the building could be obtained.

Many of the associations' names have changed today. The Underwriters National Electric Association was later to become the National Board of Fire Underwriters. The National Conference on Standard Electrical Rules was later to become the National Electrical Code Committee. AIEE was later to become the Institute of Electrical and Electronics Engineers (IEEE). The National Electric Light Association was later to become the Electric Light and Power Group.

The applications of grounding and the details of accomplishing it are many. The grounding section of the 1940 Code (Article 250) occupies over 18 pages. Curiously enough, grounding has always had a controversial aspect.

In the first (1897) National Electrical Code®, as in the previous "Rules and Requirements" issued by the Underwriters' National Electric Association, the only mandatory grounding requirement was that for lightning arresters or protectors. There couldn't be much argument about that, since from its very nature a protector or lightning arrester must be provided with a path to earth.

In the July 1, 1894, issue of "Rules for Safe Wiring," there were included directions for installing communication circuit protectors, which covered the following: (1) The location of protector; (2) Type of wiring to be used from the last outside support to the protector; (3) Conditions which grounding conductor must comply with and (4) Means for connecting grounding conductor to earth.

These provisions were carried over into the first (1897) edition of the National Electrical Code®. In addition, the frames of generators and motors were required to be grounded when, and only when, it was impracticable to insulate them.

The fireworks began about three years later, when the question of grounding secondary distribution circuits was precipitated.

Circuit grounding.

Consider for a moment what the grounding of a secondary circuit involves. Take the simple case of a two-wire, 120-volt circuit fed by a transformer. The transformer keeps the difference of potential between the two wires of the circuit at 120 volts, but, in the absence of grounding, the potential of the circuit with respect to earth is indefinite. If there is no foreign electrical influence, the average potential of the two conductors will hover around zero or earth potential, but if one of the conductors becomes crossed with a foreign wire of high potential to earth, it will immediately become charged to the same high potential (assuming that no breakdown of insulation occurs). This brings about a hazardous state of affairs, for if a person in contact with earth or with any grounded metal should touch the charged conductor, he would be liable to receive a dangerous shock. Such hazardous condition might be produced by an accidental cross with a foreign wire or by breakdown of insulation of the transformer in such way as to impress potential from the primary circuit upon a secondary circuit conductor.

As a safeguard against such hazard, it was proposed that secondary circuits be grounded, i.e., that one conductor, in case of a two-wire circuit, or the neutral conductor, in case of a three-wire circuit, be connected to earth, thus assuring that no conductor of such grounded secondary circuit could have a potential to earth greater than the normal circuit voltage.

Such grounding seemed all right for cases where the normal secondary voltage was not in itself hazardous, but as grounding increases the liability of shocks at normal circuit voltage, it did not look so good for circuits of more than about 150 volts. For example, a 220/440-volt secondary, if grounded, would increase the chance of a 220-volt shock, which of itself would involve a serious hazard. On the other hand, the grounding would protect against the probably much higher voltages due to accidental crosses with foreign wires or to transformer breakdowns. So, in this case, it was a question of balancing the increase in hazard from 220-volt shocks against the decrease in hazard from shocks at higher voltages. The 220/440-volt secondary circuit was a borderline case, some experts favored grounding while others opposed it.

As this was a technical electrical question, the Underwriters, who at this period were in control of Code revisions, wisely declined to put a mandatory circuit-grounding rule into the Code until the electrical people could agree on what the rule should be, all the more so as such a rule was considered to be more concerned with life hazard than with fire hazard. Differences of opinion on the part of electrical experts as to the circuit voltage below which grounding ought to be made mandatory produced a stalemate which persisted for several years.

Discussions of electrical circuit grounding go back to the beginnings of the ac distribution system. In England, as well as in this country, such grounding (our British cousins call it "earthing") was proposed. Thus Killingsworth Hedges said, in 1889:[7]

"One precaution is to earth the secondary circuit. Another is to connect one or both leads to a safety appliance which would automatically divert any excess current to earth and at the same time shut off the supply." Prof. Elihu Thomson, in 1890, suggested grounding the transformer secondary winding or surrounding it with a grounded sheath. As alternatives, he suggested cutting off the secondary by automatic means or grounding it automatically by film cutouts when excessive voltage appeared. The placing of a grounded metal sheath between the primary and secondary windings of a transformer was called by the English, "Kent's sheath," although the idea had its origin in America. Thomson regarded this device as very effective for securing a safety grounding connection for the secondary circuit when there was a leak from the primary. Discussing this matter at an NELA convention, he said:[1]

"In the ac system with low voltage secondary, good workmanship is all that is necessary to abolish risks if we leave out the high voltage primary which must be thoroughly insulated. Even some static capacity which may give severe if not fatal shock is to be looked out for. The writer early recognized this danger and felt it must be provided against. Hence came the expedients of grounding the secondary or surrounding it with a grounded sheath or cutting off the secondary absolutely by automatic means."

But the grounding of secondary circuits had plenty of opposition in the early days. Woodbury, in giving his suggestions about methods for reducing the electrical fire hazard, considered "it undesirable to permanently ground one wire of the secondary circuit."[2]

On March 10, 1892, the New York Board of Fire Underwriters issued a pamphlet report on "Grounding of Electric Wires," prepared by its Committee on Police and Origin of Fires. At a special meeting of the board, held on that date, it was voted that this report be printed and circulated to the members, together with certain parts of a report by Professor Henry Morton, and a reply thereto by the Edison Electric Illuminating Company of New York.

The electrical world announced:

"The New York Board of Fire Underwriters has condemned the practice of grounding the neutral as dangerous and orders it to be stopped.

"There can be no doubt that the practice of grounding the neutral is not as safe as a completely insulated system. To change it now, however, may require much work and, therefore, may interfere somewhat with the lighting. It is a case in which prevention would have been better than cure."

The following resolutions were also adopted by the board and circulated as aforesaid:

"RESOLVED, that the Committee on Police and Origin of Fires be and they are hereby directed to require all electric companies furnishing current for power and lights (to parties that have received certificates of approval of electric equipments from this Board), to make regular weekly reports of the tests of their currents, as called for in the requirements of this Board, and in the event of failure on the part of the Electric Companies so to do, then the Superintendent of the Survey Department is hereby directed to decline to grant certificates, or make inspections of equipments supplied with current by said delinquent companies, and further.

"RESOLVED, that said Committee give notice to all Electric Companies receiving certificates from this Board that the intentional grounding of any portion of the equipment is a violation of the rules of this Board and such a practice must be discontinued.

"This Board will refuse certificates of approval to all companies who do not comply with this rule, and further

"RESOLVED that the committee require of Electric Companies that all grounds now existing on electric circuits shall be removed on or before October 1st, 1892."

These resolutions were promulgated over the name of William Del. Boughton, chairman of the board.

The extracts from Professor Morton's report on fire hazards from grounding of electric wires, particularly the neutral wire in the Edison system, included the following: "The grounding of the middle wire, in my opinion, decidedly increases the fire risk, for the following reasons:

"First. If all the wires are insulated, then two ground contacts must occur in order that fire should be produced by the contact of a conductor with the gas, water or steam pipe, or other conducting substance connected with the ground. If, on the other hand, the middle wire is "grounded," then every gas, water or steam pipe becomes in fact a 'live wire,' contact with which results in a current, only limited in amount by the capacity of the conductors (including fusible catches and the like) between the point of contact and the general network of supply wires.

"Second. It renders entirely impossible any testing by which from time to time the insulation of the systems can be watched and measured.

"As a result when a ground occurs, unless it attracts attention in its own neighborhood or is enormous in amount, it can go on indefinitely without being discovered, while it may be doing mischief all the time, whereas with an insulated system, a ground will instantly show itself by the proper instruments at the central station or elsewhere, and at the same time will be harmless until another ground, on the wire of opposite polarity, is developed.

"If the middle wire of the Edison system is grounded the entire system is then in such connection with the earth that the contract of either positive or negative conductors with water, gas or steam pipe, or iron parts of a building or other conductor, will establish a connection through which a large current will flow."

Commenting on the foregoing, R. R. Bowker, first vice president of the Edison Electric Illuminating Company of New York, said:[13]

"In view of the fact that many of the questions at issue are comparatively new to members of your board, we have thought it best to make a general explanation of the nature of the three-wire system, and of the practice involved in 'grounding the neutral,' as is done in the accompanying statement, which while it does not take up the leading questions of the report in regular order, does, it will be found from the references, present a direct and sufficient answer to each of these questions. The careful and specific attention of each member of your Board is requested for this statement as the question involved is one of such large and increasing importance.

"We would emphasize to you that our street system is in no sense dependent on the practice of grounding the neutral and that contrary to the inference of Professor Morton's report, no copper or other element of cost is saved with that practice in view; the motive of the practice is not commercial but precautionary. We remind you that the practice is in vogue in other cities with the approval of Underwriters. We would also beg to correct the misapprehension that there is any grounding of the neutral wire in houses, or any decrease of insulating precautions because of that practice.

"We ask your Board for such modifications of your rules as will recognize the features of the three-wire system and provide for the practice thus adopted. We beg herewith, however, to say that should the Board, in full view of the protest of this company, decide to take the responsibility, with the insuring community and before the public, of directing this company to discontinue the practice, the superintendents of the company will have instructions to prepare to disconnect the neutral grounds at the junction boxes throughout this city, which would require, however, considerable time, and would not, we believe, produce the result desired of giving opportunities for insulating tests on lines of laboratory practice and of special circuit systems. We would, moreover, enter definite protest against the overruling of this company by your board in a question of practice which we believe can best be determined by working electrical engineers familiar with the needs and conditions of actual station practice."

A statement submitted to the New York Board of Fire Underwriters by the Edison Electric Illuminating Company contained the following:

"In the Edison 3-wire system, two dynamos are coupled together both in series by a short copper conductor connecting the negative side of one with the positive side of the other. If each is running at 100 volts the product of the two thus in series is at 200 volts. The positive conductor runs from the outer side of one dynamo; the difference of potential between the positive and negative wire is therefore 200 volts. The middle or neutral wire runs out from the short connecting link between the two dynamos and like the positive and negative wire, is absolutely continuous from the dynamo in the station to the extreme point of every house installation. Incandescent lamps are placed 'in multiple' between the neutral wire and the positive conductor and between the neutral wire and the negative conductor. In each house and throughout the system lamps are 'balanced' as evenly as possible between the two sides of the system - that is, when a hundred lamps are placed in a house, it is the endeavor to put fifty of these on the positive side and fifty on the negative side of the system, so placed that the actual burning is about the same on either side. When one thousand lamps are burning on the positive side and one thousand lamps on the negative side of the system, the neutral wire brings back no current whatever to the dynamo at the station. The current at 200 volts flows through the lamp between the positive conductor and the neutral wire, giving this 100 volts, and thence through the lamp between the middle wire and the negative conductor giving that the other 100 volts, and the return is all on the negative side. If five of these lamps are taken out or turned off on the positive side, the neutral wire would become a positive conductor to the extent of the current for these five lamps, or if five lamps were taken out or turned off on the negative side, the neutral wire would become a negative conductor to the extent of the current for those five lamps.

"On the New York system the neutral is grounded at the junction boxes in the streets, and not at the stations; on the Brooklyn system the grounding is at the station. Authorities differ as to which is the better course to obtain the advantages of grounding the neutral. In New York the neutral is grounded at about 250 junction boxes, something less than 1/3 of the number of junction boxes on the whole system, at a distance from the station varying from 1/4 to 1-1/2 miles. The neutral is grounded, as it were, on the other side of the junction box from the station, so that if in any possible event the current should flow through the ground it would seek the shortest course to the nearest junction box grounded and thence through the regular neutral wire to the station. That the ground is practically not used as the neutral conductor is sufficiently evidenced by the fact that no electrolysis has yet been discovered by our electricians at the points where the neutral wire is grounded, as would certainly be the case were the ground connection used to any extent.

"The neutral wire is not in any case grounded within house installations or in any way in connection with houses.

"It should be clearly recognized moreover that it is not practicable on any large system of electrical conductors, whether overhead or underground, especially so large as is required for the city of New York, that absolute insulation, such as is possible in laboratory tests, can be obtained or maintained. There is no known material which is an absolute insulator and even electrical appliances and adjoining woodwork are themselves conductors to an infinitesimal degree. In any known system of considerable aggregate length, while no single part can be definitely criticized, the mean level of insulation of the system is necessarily brought into comparatively close association with the ground by numerous small leaks, each of individual importance but all together making a considerable aggregate.

"I desire to make perfectly clear that if any given system could be made and kept absolutely free from leakage and best electrical authorities, including Edison himself, would agree that the balance of advantage would be found in keeping the system in that condition and not in grounding the neutral. But the difference between laboratory tests and actual working conditions is such as to make their practical conclusions almost contradictory and we have the authority of Edison's chief electrician for the assertion that 'in endeavoring to attain the perfection of safety' the report 'advocates a course of procedure which would introduce greater risk and defeat its own purpose.'

"It would seem scarcely necessary after these remarks to reply more specifically to Professor Morton's position that there should be no difference between the rules made to apply to the two-wire and three-wire systems and no reason to alter previous rules of the Board to recognize or apply to the present situation. If Rule 40, which has often been quoted in this connection, is understood by your board to apply to the grounded neutral of the Edison three-wire system, it should certainly be modified. As a matter of fact, having been adopted before the three-wire system came into operation, this company asserts that it does not so apply and in this respect no alteration would then be needed.

"It is the general opinion of the Edison interest that, while absolute insulation, if it can be had and where it can be had, is preferable, the advantages of grounding the neutral under certain conditions and particularly on large systems, are such as to make that practice in those cases the best working method particularly as a precaution against fire risk."

Discussion at national conference.

At the meeting of the National Conference on Standard Electrical Rules in March 1896, Professor Kennelly suggested a rule requiring "that the secondary coil of a converter should be grounded at its center." This led to the following discussion:

James I. Ayer:

"Relative to the question of Mr. Kennelly as to the grounding of the secondary coil of the converter at its center, it seems to me that is putting a severe strain on the insulation. I assume a low resistance ground is meant, so that in event of a ground occurring on the line and a breaking down of the insulation of the transformer the current would pass through so as to destroy the converter and cut out the primary current from the secondary. There is no doubt about the advisability of getting some sort of transformer protection, especially on the primary lines where 2000 volts and upwards are transformed directly to the house service low potential.

"In a discussion of this question some time last spring, Professor Puffer recommended the same thing. I suggested carrying out a practice which I adopted years ago in a good many cases of using a magnetic cutout. This magnetic cutout was made with two solenoids, each of which have one connection to one of the secondary lines, one on one side and one on the other, the other terminal being connected to ground. These solenoids are of very high resistance so that the leakage of a 50-volt current amounts to 8 watts and for 110 volts to about 11 watts. This gives absolute protection. Any rise of pressure on the secondary mains above 500 volts would operate the circuit breaker. It is very positive. It rather straddles the question of grounding secondaries or grounding house mains by grounding them through a fixed high resistance ground connection. It gives an absolute, positive-acting device, not necessarily one that burns up transformers. In fact, it leaves the transformers to take care of itself but cuts out the primary from the secondary main.

"This is a matter to which I have given a great deal of thought. I found in the early days of the use of transformers the necessity of some such protection where there are many of them connected with the line. Thunderstorms would readily break them down and create punctures which would give a connection from one side of the transformer to one side of the secondary circuit. The lights would burn and things continue on apparently all right. Transformers do break down with the high potential being used and I believe the right thing to recommend, and in time to demand, is the proper protection for isolated transformers, to give you protection between the secondary and primary, but the question of grounding secondary mains and low resistances IS a very serious one and should receive a very full discussion before action is taken."

Prof. Kennelly:

"I think the matter is so important that it is worth while to give it a few moments. I should be sorry to advocate anything which would unduly harass the manufacturers, but I believe that I make this suggested recommendation in the interests of the manufacturers just as much as in the interests of the parties who receive lighting from electric systems.

"If we exclude lightning, there can be no objection to grounding the secondary because it is obvious that the system should be capable of standing the working pressure. Then no one can fairly object to grounding the center of the secondary on the score of danger of puncture because, if the transformers are going to be so weakly insulated that grounding the secondary is going to injure them, those transformers had better be replaced. It may not be necessary, however, to actually ground the secondaries. A simple protective device that will ground the secondary if a cross develops would be preferable, and if Mr. Ayer's device is simple and practical, its adoption would seem worthy of consideration. A cross between the primary and coils of a transformer is, we know, of very rare occurrence, yet we all know that no one would be justified in permitting a person standing, say on a cellar floor, to handle the socket of a lamp if the operation involved making contact with one of the secondary conductors, unless the transformer had just been tested free from leakage or cross. If, however, the secondary coil of the transformer were grounded permanently, or protected in an equivalent degree there would be no such danger from accidental contact."

Mr. Ayer:

"In considering this question of the protection of the transformers, the practical conditions to be met with have got to be well understood. You take the case of distribution with the overhead system, where there are any material number of transformers in a network covering a large area, there is no such thing as getting an insulated circuit. You have got a grounded circuit, always.

"There has been a great deal of discussion especially among the Edison Companies and Underwriters, as to the wisdom of grounding the third wire of the Edison system. That has been a very fruitful topic of discussion. I think it is useless to try to enumerate the details or discuss the pros and cons of this here.

"In a resume of the English rules, you see that they contemplate at all times an automatic device in the secondary circuit, or in the primary wire, if not in the secondary, which will automatically cut off the primary current from the transformer or the secondary from the house. That is the English practice. They call for some intervening device which shall open the secondary circuit in the event of the primary forming a connection with high tension. It is a simple thing to make. I have one in my residence today and have had for five years, which has been in service all the time. The thing is operated with a leakage of 500 volts from the primary line to the secondary. It introduces a high resistance ground on the secondary and a leakage on the secondary of about 8 watts. I believe the solution of the problem of the protection of the secondary coil is the introduction of this intermediate automatic device. Devices have been made in England operated by clock work, fuses, etc., but not similar to this to my knowledge."

The protective device for secondary circuits, which Ayer thus commended, did not meet with much favor and the solution of the problem was long delayed; indeed, it has not been solved even yet in all its aspects.

The Underwriters were slow to reverse their stand against allowing any permanent grounds on electric circuits.

The electric companies, on the other hand, were generally in favor of grounding the neutral wire. W.L.R. Emmet of the General Electric Company said, in 1899:

"The permanent grounding of transformer secondaries is prohibited by the fire underwriters' rules, which pro prohibition is generally respected throughout the country, so that most secondary circuits are entirely unprotected."[15]

AIEE talks it over.

In 1899, Dr. Cary T. Hutchinson presented an AIEE paper on the protection of secondary circuits, in the course of which he heartily endorsed the practice of grounding and reproved the Underwriters for their stand against it saying,[8]

"The need of some effective method for protecting secondary circuits is well recognized yet there is at present probably no device for the purpose that is thoroughly trustworthy and efficient in all circumstances. The means that have been employed for this object may be classified under three general heads: (1) Devices intended to ground, short-circuit or open-circuit the secondary circuit when subjected to abnormal difference of potential; (2) grounded metallic shields imposed between the primary and secondary coils of the transformer; (3) permanent grounding of the secondary system.

"The third method, grounding the secondary permanently, is the only sure way to prevent the potential above earth of the secondary system rising above the voltage of the circuit.

"The grounding of the secondary circuit thus absolutely insures the safety of the circuit as regards abnormal pressure.

"It would seem that a remedy as simple as this, one generally applicable, would have been applied in all cases but the fact is that it has attained a limited use only. The reason for this state of affairs is the refusal of the Board of Fire Underwriters to authorize the practice of grounding any part of a circuit carrying current. The Underwriters take the position that to ground one side of a current

brings about an increased liability to fire, because the full voltage of the circuit continually acts upon the insulation of the circuit, instead of one-half of this voltage as in an insulated system, and because one accidental ground on the insulated side may cause a fire.

"The Association of Edison Illuminating Companies has had a committee on grounding the neutral since 1890. The recommendations of this committee being based on extensive experience are entitled to great weight. Their recommendation of uniformity being the greatest safety was assured by the practice of grounding the neutral. This is the practice of nearly all the large Edison Companies as is well known. In some cases the companies have been forced to this position because they were not able to free the neutral from ground. In other cases they have deliberately adopted this as the best remedy for many troubles due to operation dangers from fire and to abnormal pressures.

"It is notorious fact that the systems in nearly all of the large cities, particularly the Edison system, are grounded. The neutrals in most cases are permanently grounded at the junction boxes. It is equally well known that in several of the large ac distributing systems the neutral wires are grounded. The authorities of the Board of Underwriters certainly should, and probably do know, these facts. To keep rules in print prohibiting such practices under the circumstances must weaken their authority and lessen the respect for all their rules.

"I therefore offer the following resolutions:

"RESOLVED that the AIEE favors a rule permitting the grounding of one wire of every low potential consumption system, and

"RESOLVED that the committee of the AIEE on the National Electrical Code is hereby requested to confer with the Electrical Committee of the Underwriters National Electric Association in relation to recommending to the National Board of Fire Underwriters the passage of a rule permitting the permanent grounding of one wire of such systems under suitable restrictions."

Hutchinson's resolution was passed, after discussion, some of which is noted below: Professor Elihu Thomson: "I endorse everything that is said and I hope some action will be taken by a body so influential as this, toward getting some such thing adopted."

C. M. Goddard:

"I should like to second the resolution and for several reasons; not, however, because the Underwriters are fully convinced as yet that one of their rules, almost as unchangeable as the laws of the

Medes and Persians, that no circuit should be permanently grounded, has been wrong all the time. I am not convinced of that fact but because we are delighted to have the institute take up any subject in relation to our rules and try to help us arrive at a right solution.

"I think it very probable that this grounding of the secondary may be the best solution but the Underwriters are always conservative in such matters.

"We would be delighted to have you help us in making our rules. We modified this rule last year at the request of the Institute and we will certainly modify it further if you will only come to us and show us that you are right, for we have a great deal of respect for the opinions of the Institute."

Professor W. E. Goldsborough:

"I hardly feel like saying anything in opposition but I am very much interested in the subject and all my experience has tended to lead me to feel it is not best to ground the secondary."

Professor W. L. Puffer:

"I would carry grounding to such an extent that I would compel the absolute grounding immediately at the point of entrance in buildings of every system (except series arcs which ought not to be allowed in buildings) and including such things as water pipes and gas pipes."

While the consensus of this discussion was decidedly in favor of grounding the secondaries, there was some adverse opinion and the Underwriters seem to have continued other opposition to the practice, although the Code, i.e., the 1897 and 1899 editions, did not specifically prohibit it.

Father defers to mother.

The Electrical World and Engineer, reporting the December, 1900, meeting of the Electrical Committee, said:[11]

"The motion to change the limit of low tension circuits from 300 to 550 volts was carried at the first session. A long discussion took place on the advisability of grounding one of the wires in low tension circuits in which numerous authorities were quoted and prominent electrical engineers spoke in its favor. The resolution was finally adopted to allow the grounding of one wire in every low potential circuit but making it optional with the installing engineer."

In accordance with this decision, the 1901 edition of the Code contained the following section:

"13A. Grounding Low Potential Circuits.

The grounding of low potential circuits under the following regulations is only allowed when such circuits are so arranged that under normal conditions of service there will be no passage of current over the ground wire.

"Direct-Current 3-Wire Systems:

a. Neutral wire may be grounded and when grounded the following rules must be complied with:

1. Must be grounded at the central station on a metal plate buried in coke beneath permanent moisture level and also through all available underground water and gas-pipe systems.
2. In underground systems the neutral wire must also be grounded at each distributing box through the box.
3. In overhead systems the neutral wire must be grounded every 500 feet, as provided in Sections c, e, f, and g.

Inspection departments having jurisdiction may require grounding if they deem it necessary.

Two-wire direct-current systems having no accessible neutral point are not to be grounded.

"Alternating-Current Secondary Systems

b. The neutral point of transformers or the neutral wire of distributing systems may be grounded and when grounded, the following rules must be complied with:

"1. Transformers feeding 2-wire systems must be grounded at the center of the secondary coils, as provided in sections d, e, f and g.
"2. Transformers feeding systems with a neutral wire must have the neutral wire grounded as provided in sections d, e, f and g, at the transformer and at least every 250 feet for overhead systems and every 500 feet for underground systems.

"Inspection departments having jurisdiction may *require* grounding if they deem it necessary."

Thus, by making the practice permissive, the Underwriters began a strategic retreat from their early position against the grounding of electrical circuits.

At the Electrical Committee meeting in December, 1901, the following proposal was considered:[3]

"A letter was also referred for conference with the AIEE Committee, asking for modification of the rules with regard to grounding low potential circuits so that the ground might be on one side of the circuit instead of at the neutral point of the transformer."

The Underwriters held to their position that the grounding of secondaries was a question for the electrical people to decide. A special subcommittee on the subject reported to the Electrical Committee meeting of December 9, 1902:[9]

"It was also agreed that Rule 13A-b should be amended to read as follows:

"(b) Transformer secondaries of distributing systems should preferably be grounded and when grounded the following rules must be applied:

"1. The grounding must be made at the neutral point or wire, whenever a neutral point or wire is accessible.

"2. When no neutral point or wire is accessible, the maximum difference of potential from the grounded point of the secondary circuit to any other point there of must not exceed 250 volts.

"3. To be identical with the present clause marked (2)." Here was a proposal to go further and take a position favoring circuit grounding. The meeting declined to either approve or reject this recommendation but voted to submit it to the Board of Directors of the AIEE and to the Underwriters' National Electric Association.

This recommendation to change the permissive "may be grounded" to "should preferably be grounded," was evidently accepted by the AIEE, for the rule in the 1903 Code was substantially the same as the subcommittee had recommended, namely:

"b. Transformer secondaries of distributing systems should preferably be grounded, and when grounded, the following rules must be complied with:

1. The grounding must be made at the neutral point or wire, whenever a neutral point or wire is accessible.

2. When no neutral point or wire is accessible, one side of the secondary circuit may be grounded, provided the maximum difference of potential between the grounded point and any other point in the circuit does not exceed 250 volts.

3. The ground connection must be at the transformer as provided in sections d, e, f, g and when transformers feed systems with a neutral wire, the neutral wire must also be grounded at least every 250 feet for overhead systems, and every 500 feet for underground systems. "Inspection Departments having jurisdiction may require grounding if they deem it necessary."

This rule in the 1903 Code represented quite a step away from the old position prohibiting circuit grounding, but still it was unsatisfactory in that it left the question open as to whether secondaries should be grounded or not. There was a substantial body of opinion in favor of making the grounding mandatory under certain conditions, but there were differences of opinion as to what the conditions were which warranted making grounding mandatory. It was generally agreed that where the voltage between conductor and ground did not exceed 150 volts, safety was promoted by grounding, but where the voltage was from 150 volts to 250 volts, the advantage of grounding was in a doubtful region. For voltages to ground over 250, the consensus of expert opinion was against grounding.

At the Electrical Committee meeting on December 2-3, 1903, a suggestion that the grounding of secondary neutrals be made compulsory was referred to a committee of three to be appointed by the chair, to report back.

More discussion.

At an AIEE meeting on December 19, 1903, H. G. Stott took up the discussion, saying:

"Some years ago a committee of the Institute met with the Board of Fire Underwriters and a rule was passed making it permissible to ground the neutral at the central point of the transformer or at one side of the transformer. The grounding of any part of the secondary network, especially the neutral point, is the best form of protection obtainable because in case of a cross with an electric light wire carrying perhaps 8,000 volts, the pressure of the wire is immediately reduced to the pressure of the circuit as the carrying capacity of the secondary network is greatly in excess of the amount of current passing through the average arc light circuit."

Discussion of this question continued. It was one of the subjects considered at the meeting of the National Conference on April 21, 1905. On that occasion, Dr. F. A. C. Perrine (formerly Professor of Electrical Engineering at Stanford University) stressed—perhaps overstressed—the opposing effects of grounding on the fire and life hazards, saying:

"For example, we have heard a good deal of the rules for grounding secondaries. We know the Underwriters have admitted that, at the request of the electrical interests, in spite of the fact that if anything it increases the fire hazard, decreases the life hazard. Now there is a condition in which the Underwriters have acceded to the request of the electrical interests but it is a condition in which the fire hazard risk is distinctly opposed to the life hazard risk."

Benallack confirmed the influence exerted on the Underwriters by the electrical people:[4]

"Other speakers had called attention to the importance of grounding secondary circuits as a good way, and in fact the only way, of making these circuits safe from the leakage of currents from the primary circuits. The Underwriters had been slow to permit this at first and it was largely due to the central station men who appreciated the danger that the practice was permitted and finally positively recommended by the Underwriters."

At the National meeting on December 4, 1905, the subject of grounding secondaries was discussed further. H.C. Wirt remarked:

"The most important rule here is the rule on grounding of the secondary to make it compulsory." After discussion, the following resolutions were adopted:

"RESOLVED, that it is the sense of this Conference that the grounding of low tension secondary ac systems be strongly recommended in all cases wherever reliable grounding connections can be secured. "RESOLVED, that in cases of three-wire systems with grounded neutral, solid connections without fuse be permitted on the neutral wire.

"RESOLVED, that it is the sense of this Conference that the grounding of secondary ac systems inside of buildings to water pipes, provided such connection is made at the nearest point to cellar wall on water pipes and outside of meter, is not only safe but places no additional burden or menace on such water pipes."

Woodbury referred to this meeting in his Annual Report for 1905, to H. V. Hayes:

"At the second meeting of the Conference on December 4, results of importance to the electrical interests were adopted in regard to grounding of secondaries on ac systems and also the resolution that grounding did not furnish any hazard or menace to water pipes to which they were attached. Although these results were filed by others yet I was concerned in their operation, particularly in relation to the last clause, on account of difficulties which some of the operating companies had experienced through claims of town officials for reimbursement for damage to water mains by lightning, alleging that the grounding of telephone lines to water pipes was the cause."

However, the Electrical Committee at its December 1905, meeting did not adopt the National Conference resolutions, as no formal endorsement of them by the AIEE or NELA had been received. The grounding of low-potential circuits was discussed but no changes in the rule were made and the matter was referred back to the subcommittee for further consideration. Nor was any such endorsement, or any further recommendation as to the grounding rule, received from the electrical people in 1906. Accordingly, in the 1907 edition of the Code, rule 13A remained practically unchanged

from the previous edition. The year 1908, however, saw some progress toward a mandatory grounding rule for the Code. At the Electrical Committee meeting on March 25, the Committee on Rule 13A reported:

"The sense of the meeting was that, when suitable rules are prepared, the grounding of secondary systems shall be made mandatory."

At the meeting of the National Conference on March 27, 1908, C.M. Goddard made these remarks: "The Underwriters have all along felt that it (grounding) was a rule which did not very seriously affect the fire hazard but I think the more it has been looked into, that the Underwriters have perhaps decided that there are certain fire hazards guarded against by the grounding although, of course, we all know that it is principally for guarding against life hazard."

"It is apparent that when the rule has been whipped into shape so that it will be satisfactory to all, it shall be made mandatory by the Underwriters instead of being simply permissive.

"Now the Underwriters want to be very sure that it is the desire of the electrical interests as a whole, both central station men and the engineering branch, also the municipal departments. We want to be very sure that we are not making the rule mandatory against their wishes, I should think the Underwriters would be perfectly satisfied to leave the rule as it stands."

It was voted to be the sense of the meeting that the grounding rule be made mandatory; also that instructions be given for making the grounds.

At the 17th annual meeting of the Electrical Committee, held in New York on March 24 and 25, 1909, the Committee on Rule 13A reported:[6]

"Your committee appointed to consider changes in grounding rule, and question as to whether it should be made mandatory, begs to report as follows:

"First—As we are asked only to make this rule mandatory up to and including 150 volts for alternating current secondary circuits, and to prohibit grounding of such circuits carrying in excess of 150 volts, it is evident that this suggestion cannot be based on the question of fire hazard because the fire hazard must certainly be as great from an ungrounded circuit in excess of 150 volts as from such circuit carrying not over 150 volts.

"Second—Your committee believes that the Underwriters have no right to enforce a rule which they cannot defend on the ground that it is adopted for the purpose of lessening the fire hazard.

"Third—Your committee believes that the Underwriters should use all of their moral influence in supporting any rule which may seem advisable in order to lessen the life hazard.

"Your committee would, therefore, recommend no change in the present rule but suggest that the rule, if made mandatory, might be followed in the rules of the National Board of Fire Underwriters for the government of insurance inspectors by a note saying, that in order to lessen the life hazard, compliance with the rule is recommended, but as the rule is not adopted for the purpose of lessening the fire hazard, no penalty could be enforced by the Underwriters for failure to observe its provisions.

"Your committee regrets that they have received no information as to the position of the American Institute of Electrical Engineers or the Associated Edison Illuminating Companies in regard to the subject."

This report was adopted by the meeting and the grounding section was again left unchanged in the 1909 Code.

In its annual report for 1909, the Board of Directors of AIEE included this passage:[14]

"The Code Committee has held several meetings, the most important matter considered being in connection with Rule 13A of the Board of Fire Underwriters. This rule deals with the grounding of secondary circuits. The committee agreed that the rules should be made mandatory up to and including 150 volts between any wire and ground, and optional up to, and including, 250 volts between any wire and ground. This resolution was handed to the president of the National Conference on Standard Electrical Rules. Final action on the resolution was deferred pending an arrangement for a meeting of the Code Committee with a committee of the National Board of Fire Underwriters."

So, in the 1911 Code the grounding rule, which in revision became Rule 15, was kept about as it had been, to wit:

"15. Grounding Low Potential Circuits

"Alternating-Current Secondary Systems

b. Transformer secondaries of distributing systems should preferably be grounded, and when grounded the following rules must be complied with:

1. The grounding must be made at the neutral point or wire, whenever a neutral point or wire is accessible.

2. When no neutral point or wire is accessible, one side of the secondary circuit may be grounded, provided the maximum difference of potential between the grounded point and any other point in the circuit does not exceed 250 volts.

3. The ground connection must be at the transformers or on the individual service as provided in sections c to g, and when

transformers feed systems with a neutral wire, the neutral wire must also be grounded at least every 500 feet."

"Inspection Departments having jurisdiction may require grounding if they deem it necessary."

But the time was now approaching when the grounding of low-tension secondaries was to be made mandatory.

In its report for 1911, the Board of Directors of the AIEE included the following passage:[10]

"The Code Committee through its chairman, represented the institute at the annual meeting of the National Board of Fire Underwriters held at New York on March 20-21, 1911. The only matter of interest to the institute taken up at this meeting was the grounding of secondaries and the work of the institute's representative resulted in the passing of a resolution by the underwriters conference enforcing the practice of the grounding of secondaries and recommending that municipalities and lighting companies make such a rule mandatory with the further resolution that the institute use its efforts to bring about an agreement with the NELA in the matter of grounding secondaries up to 250 volts instead of at 150 volts, the present adopted standard of the Association."

In March 1911, the Electrical Committee held its first meeting as the adopted child of the National Fire Protection Association. The following is an extract from a report of that meeting:[12]

"The following resolution, presented by the Committee on Rule 13A, was adopted:

"WHEREAS—The grounding of secondary alternating-current circuits having a normal difference of potential of not over 150 volts does effectually eliminate the life hazard of such circuits due to their accidental contact with circuits of a dangerous potential without introducing any hazard due to grounding, and

"WHEREAS—There is a difference of opinion as to whether the grounding of such circuits having a normal difference of potential of over 150 volts introduces hazards due to such grounding, and

"WHEREAS—Such grounding does not increase but rather tends to decrease, the fire hazard, therefore be it

"RESOLVED—That municipal departments are urged to make the grounding of secondary circuits up to 150 volts mandatory as a necessary safeguard to life; and

"RESOLVED—That Underwriters Inspection Departments be urged to recommend at all times such grounding as a proper and desirable precaution which introduces no fire hazard, and

"RESOLVED—That the National Electric Light Association be urged to see that all of its member companies be brought to realize the necessity of such grounding for the protection of their customers, and

"RESOLVED—That departments in charge of water works be urged to allow the attaching of such ground wires to their piping system in the full confidence that the integrity of such piping systems will in no way be affected, whatever may be the normal voltage, and

"RESOLVED—That all concerned give careful thought to other methods of obtaining satisfactory grounds wherever metallic water pipes are not available, to the end that this most necessary precaution of grounded secondary circuits may be available in all localities and under all conditions, and

"RESOLVED—That the American Institute of Electrical Engineers be urged to use its best endeavors to harmonize the present difference of opinion as to the limit of voltage at which grounding ceases to be desirable by determining a limit which shall meet with general approval."

During the year following, the American Institute of Electrical Engineers joined with other prominent national electrical societies in a final effort to secure a mandatory requirement for the grounding of low-potential secondary circuits. The AIEE Board of Directors' report for 1912, contained this passage:

"The Code Committee held a meeting on March 12, 1912, with representatives of the National Electric Light Association, Association of Edison Illuminating Companies, the National Inspectors Association, and concurred in a joint recommendation to the NFPA in regard to the grounding of secondaries."

With this substantial backing, the Electrical Committee was encouraged to approve, at its meeting on March 26 and 27, 1913, the report of its Committee on Rule 15, which made "the grounding of transformer secondary circuits obligatory."

Circuit Grounding Made Compulsory

So, at last, in the 1913 Code, a mandatory grounding rule was included, reading as follows:

"15. Grounding Low-Potential Circuits

"Alternating-Current Secondary Systems b. Transformer secondaries of distributing systems (except when supplied from private industrial power or lighting plants when the primary voltage does not exceed 550 volts) must be grounded, provided the maximum difference of potential between the grounded point and any other point in the circuit does not exceed 150 volts and may be grounded when the maximum difference of potential between the grounded point and any other point in the circuit exceeds 150 volts. In either case the following rules must be complied with:

"1. The grounding must be made at the neutral point or wire, whenever a neutral point or wire is accessible.

2. When no neutral point or wire is accessible, one side of the secondary circuit must be grounded.

3. The ground connection must be at the transformers or on the individual service as provided in sections c to g, and when transformers feed systems with a neutral wire, the neutral wire must also be grounded at least every 500 feet."

This rule has not been changed much, in substance, since that time. The main difference is that in the 1940 Code, grounding for voltages from 150 to 300 is recommended, instead of being merely permitted. This change first appeared in the 1923 edition.

References

1. *"Safety and Safety Devices in Electrical Installations" by Professor Elihu Thomson, Electrical World, Feb. 22, 1890.*
2. *Address at Cornell University by C. J. H. Woodbury, Electrical Engineer, November 18, 1891.*
3. *Electrical World and Engineer, December 14, 1901.*
4. *"The Electrical Fire Hazard" by William T. Bennallack, Journal Western Society of Engineers, Chicago IL., March 10, 1905.*
5. *National Electrical Contractor, January, 1906.*
6. *National Electrical Contractor, April, 1909.*
7. *"The Fire Risks of Electric Lighting," by Killingsworth Hedges, Electrical World, October 5, 1889*
8. *"The Protection of Secondary Circuits from Fire Risk" by Dr. Cary T. Hutchinson, Trans. AIEE, Vol. XVI, 1899.*
9. *National Electrical Contractor, January, 1903.*
10. *Trans. AIEE, 1911, Part 3, P. 2591.*
11. *Electrical World and Engineer, December 15, 1900*
12. *National Electrical Contractor, April, 1911.*
13. *Electrical World, March 26, 1892.*
14. *Trans. AIEE 1909, Part 2, p. 1513.*
15. *"Means of Attaining Safety in Electrical Distribution" by W. L. R. Emmet, Electrical Review, June 7, 1899.*

Specifications to assure safety

Certain conclusions can thus be established to assist in writing specifications to cover Service and Equipment Grounding to assure safety in installations. They are as follows:

1. All ungrounded systems should have a ground detector and proper maintenance applied to avoid, as far as practicable, the occurrence of grounds on opposite phases at the same time. If proper maintenance cannot be assured for an ungrounded system then a grounded system should be installed.

2. The use of grounding electrodes at a transformer bank and at the building service provide by themselves virtually no protection in the event of a ground fault, as far as overcurrent protection is concerned. However, the lowest practical resistance for such grounds is desirable to serve the second purpose of holding the equipment as near to ground potential as possible.

3. The service ground and the equipment ground must have their common point of connection within the service equipment enclosure. If the service equipment is in the switchboard, then the common point must be within the switchboard section which houses the service equipment. A common grounding conductor to the grounding electrode must be run from that common point of connection.

4. When a transformer bank is grounded, that is, when a service is run to a building from a system which has any point grounded, the grounded conductor must be run to the service equipment whether that grounded conductor is to be used for voltage requirements or not, and that grounded conductor must be connected to the equipment grounding conductor within the service equipment enclosure.

5. When a fault occurs on any grounded system we must not assume it began as a short circuit. (phase/to/phase fault) until we have clearly indisputable evidence to the contrary, if we are to analyze the fault correctly. It is almost self-evident, when the overcurrent devices do not operate promptly that the trouble was owing to a ground fault.

Unless "effective grounding" is followed and carefully observed, we cannot expect a ground fault to clear unless the circuit is manually opened or the ground fault continues and develops into a phase to phase fault. The latter condition may require many minutes during which time the equipment grounding conductor will develop a potential to ground, which may become high enough to be dangerous. Further, we may have a disastrous fire loss. Some method of automatically clearing a ground fault in limited time is highly desirable.

Engineering specifications must be carefully written to cover service and equipment grounding and not left to chance interpretation of how to accomplish safety. It is not an uncommon practice for specifications to read: "Ground in accordance with the provisions of Article 250 of the National Electrical Code®." The engineer is not assuming his full responsibility for the safety of the installation when he writes such specifications. It is the engineer's responsibility to outline in his specifications how the job should be installed to fully comply with all three of the objectives of "effective grounding" and not leave it to chance, as is now often the case.

Ufer Ground

H.G. Ufer, in an IEEE Conference Paper, CP-61-978, describes an installation of made ground electrodes on 24 buildings in 1942, in Arizona, to meet a 5-ohm maximum value. The resistance values were checked bimonthly over an 18-year period, during which time no servicing was required.

In 1960, the maximum reading was 4.8 ohms and the minimum 2.1 ohms. The average value of the 24 installations was 3.57 ohms.

The installations used ½-inch steel reinforcing rods set in a concrete footing. They were at two locations in Arizona. The first was near Tucson, Arizona, which is normally hot and dry during most of the year and has an average annual rainfall of 10.91 inches. The soil is sand and gravel. The second location was near Flagstaff, Arizona, where the soil is clay, shale gumbo and loam with small area stratas of soft limestone. The made electrodes were used as no water piping system was available.

As a result of these installations and the 18-year test period, Mr. Ufer suggested that a No. 4 or larger copper wire be embedded in the concrete footing of a building and that test data be compiled further to verify the effectiveness. Based on this data, CMP-5 accepted a concrete-encased electrode commonly referred to as a "Ufer Ground." The concrete-encased electrode shall consist of at least 20 feet of bare copper not smaller than No. 4 AWG encased in 2 inches of concrete near the bottom of the footing or foundation.

Answers to Questions Chapter One - Seventeen

Answers to questions in Chapters 1-17.
The references included below can be found on the pages in this textbook and in the 1996 National Electrical Code® under the sections given.

Chapter 1
1. c. in textbook
2. d. Section 250-l (FPN No. 1)
3. d. Section 250-51
4. d. in textbook
5. b. in textbook
6. a. in textbook
7. b. in textbook
8. b. in textbook
9. d. in textbook

Chapter 2
1. d. Section 250-5(a)(1)
2. b. Section 250-5(a)(2)
3. a. Section 250-5(a)(3)
4. b. Section 250-5(b)(1)
5. b. Section 250-5(b)(2)
6. c. Section 250-5(b)(3)
7. d. Sections 250-5(b) and 230-22
8. b. Sections 250-5(b) and 230-30
9. c. Sections 250-5(b)(4) and 230-41
10. d. Section 250-5(b), Exception No. 1
11. b. Section 250-5(b), Exception No. 2
12. d. Section 250-5(b), Exception No. 3
13. b. Sections 250-5(b), Exception No. 4, and 668-30(a)
14. d. Section 250-5(b), Exception No. 5
15. b. Section 250-5(c)
16. a. Section 250-5(c)
17. d. in textbook
18. d. Section 250-5(d)
19. a. Section 250-5(d)
20. b. in textbook
21. c. in textbook
22. d. in textbook
23. d. in textbook
24. a. in textbook

Chapter 3
1. b. Article 100
2. a. Article 100
3. c. Article 100
4. a. Section 200-6(a)
5. c. Section 200-6(a), Exception No. 3
6. d. Section 200-6(b)
7. c. Sections 215-8, 230-56 and 384-3(e)
8. d. in textbook

Chapter 3 *continued*
9. a. in textbook
10. d. in textbook
11. b. in textbook
12. d. in textbook
13. c. in textbook
14. b. in textbook
15. d. in textbook
16. c. in textbook
17. b. in textbook

Chapter 4
1. c. Section 250-23(b)
2. d. Section 250-23(b)
3. a. Section 250-23(b)
4. b. Section 250-23(b)
5. a. Section 250-23(b)
6. c. Section 250-23(b)
7. c. Section 250-23(b), Exception No. 3
8. a. in textbook
9. d. in textbook
10. a. Section 250-23(a)
11. c. Table 250-94
12. c. Article 310, Note 3
13. d. in textbook
14. d. Section 250-5(b), Exception No. 5
15. a. Section 250-121
16. b. Section 250-121, Exception
17. d. Section 250-122, Exception

Chapter 5
1. b. Article 100
2. a. Article 100
3. d. Section 250-79(a)
4. b. Section 250-79(b)
5. d. Table 250-94
6. d. Section 250-79(d)
7. a. Section 250-79(d)
8. d. Section 250-71
9. b. Section 250-61(a)
10. c. in textbook

Chapter 6
1. b. in textbook
2. a. Article 100
3. d. Section 250-81(c)
4. c. Section 250-81(c)
5. b. Section 250-81(d)
6. a. Section 250-81(a)
7. d. Section 250-81 (a)
8. c. Section 250-81

9. d. Section 250-83
10. a. Section 250-54
11. d. Section 250-21(d)
12. b. Section 250-21(d)
13. d. Section 250-84
14. a. in textbook
15. d. in textbook
16. b. Section 250-86
17. b. Section 250-46

Chapter 7
1. b. Article 100
2. c. Section 250-94
3. a. in textbook
4. c. Table 250-94
5. c. Table 250-94
6. d. Table 250-94
7. d. Table 250-94
8. c. Section 250-71(a)
9. b. Section 250-94, Exception: a.
10. b. Section 250-94, Exception: b.
11. d. Section 250-115
12. b. Section 250-115
13. a. Section 250-92(a)
14. b. Section 250-92(a)
15. b. Section 250-92(a)

Chapter 8
1. d. Section 250-76
2. d. Section 250-113
3. d. Section 370-3, Exception Nos. 1 and 2
4. c. Section 250-114
5. b. Section 250-74, Exception No. 4
6. c. Section 250-79(f)
7. d. Section 250-80(b)
8. d. in textbook
9. c. Section 250-80(b)
10. b. Section 250-77
11. a. Section 250-78
12. c. Section 250-95
13. c. Section 250-80(a)
14. c. in textbook
15. b. Section 250-80(a)
16. d. Section 250-80(c)

Chapter 9
1. a. Article 100
2. d. in textbook
3. c. in textbook
4. a. Table 250-95
5. d. Section 250-91(b)(11) listed
6. c. Section 250-91(b)

Chapter 9 *continued*

7. c. Section 250-91(b), Exception No. 2
8. c. Section 250-92(c)(1)
9. b. Sections 250-95, Ex. No. 1 and 240-4
10. b. Section 250-58(a)

Chapter 10

1. c. Article 100
2. a. Article 100
3. b. Section 250-42(a)
4. d. Section 250-42(b)
5. d. Section 250-42(e)
6. b. Section 250-42(f), Exception No. 1
7. c. Section 250-42(f), Exception No. 2
8. b. Section 250-42(f), Exception No. 3
9. d. Section 250-42(f), Exception No. 4
10. b. Section 250-33, Exception No. 1
11. a. Section 250-43(k)
12. d. Section 250-113
13. b. Section 250-113
14. c. Section 250-32, Exception
15. d. Section 250-60
16. b. Sections 250-23(a), 215-6 and 384-20

Chapter 11

1. c. in textbook
2. b. in textbook
3. c. in textbook
4. b. in textbook
5. a. in textbook
6. b. in textbook
7. d. in textbook
8. b. in textbook
9. a. in textbook
10. b. in textbook
11. a. No. 6, in textbook
12. b. No. 2, in textbook
13. c. No. 10, in textbook
14. d. No. 2, in textbook

Chapter 12

1. c. Article 100
2. b. Section 250-26(b) and Table 250-94
3. a. Section 250-26(b)
4. c. Section 250-26(c)
5. b. Section 250-26(c)(1)
6. c. Section 250-26(c)(2)
7. a. Section 250-54
8. d. in textbook
9. a. in textbook
10. b. in textbook
11. d. in textbook

Chapter 13

1. b. Article 100
2. c. Section 250-24(a)
3. d. in textbook
4. c. in textbook
5. a. Section 250-83
6. c. Section 250-95
7. c. in textbook
8. a. Section 250-61(a)(3)
9. a. Section 250-21(d)
10. d. Section 547-8(c)

Chapter 14

1. b. Section 210-7(d)
2. a. Section 210-7(d), Exception
3. b. Section 210-8(a)(2), Exception No. 2
4. c. Section 210-8(a)(4)
5. c. Section 210-8(a)(5)
6. c. Section 210-63
7. d. Section 210-8(b)(2)
8. d. Section 210-8(b)(2)
9. b. in textbook
10. a. in textbook
11. a. in textbook
12. b. in textbook
13. a. in textbook

Chapter 15

1. b. Article 100
2. b. Section 230-95
3. c. Section 230-95, Exception Nos. 1 and 2
4. b. Section 230-95(a)
5. a. in textbook
6. c. in textbook
7. a. in textbook
8. d. Section 230-95(c)
9. b. Section 517-17
10. c. Section 215-10

Chapter 16

1. b. Section 250-46
2. c. Section 250-46
3. c. Sections 501-16(a), 502-16(a) and 503-16(a)
4. d. in textbook
5. a. Section 250-79(f)
6. b. Section 547-8(c)
7. a. Section 547-8(b)
8. d. in textbook
9. d. in textbook
10. d. Section 517-13(b)
11. d. Section 517-19(d)
12. c. Section 680-22(a)(6)

Chapter 16 *continued*

13. a. Section 680-22(b)(3)
14. a. Section 680-20(c)
15. d. Section 680-6(a)(3)
16. c. Section 680-5(c), Exception 1
17. a. Section 680-7
18. c. Section 680-5(b)
19. d. Section 680-20(a)(1)
20. d. Section 680-20(a)(1)
21. b. Section 680-20(b)(3)
22. a. Section 680-20(b)(3)
23. b. Section 680-5(a)
24. b. Section 680-5(a)

Chapter 17

1. b. in textbook
2. a. in textbook
3. b. in textbook
4. c. in textbook
5. c. in textbook
6. a. in textbook
7. c. in textbook
8. b. in textbook
9. c. in textbook
10. d. in textbook
11. a. in textbook
12. b. in textbook

Notes:

Index

International Association of Electrical Inspectors'
Major Objectives

- Formulation of standards for safe installation and use of electrical materials, devices and appliances.
- Promotion of uniform understanding and application of the National Electrical Code®, other codes, and any adopted electrical codes in other countries.
- Promotion of uniform administrative ordinances and inspection methods.
- Collection and dissemination of information relative to the safe use of electricity.
- Representation of electrical inspectors in the electrical industry, nationally and internationally.
- Cooperation with national and international organizations in the further development of the electrical industry.
- Promotion of cooperation among inspectors, inspection departments of city, county and state at national and international levels, the electrical industry and the public.

Contact IAEI membership secretary for information on our special membership categories - Section, National and International Member; Sustaining Member (Bronze, Silver, Gold or Platinum); and Inspection Agency Member.

MEMBERSHIP APPLICATION

for
Inspector Member
Associate Member
Student Member

Annual Membership Dues $40
Students $30

Services from the IAEI for you and your organization include all this and more:

- Subscription to *IAEI News* - Bimonthly magazine with latest news on Code changes
- Local Chapter or Division meetings - New products, programs, national experts answer questions on local problems
- Section meetings to promote education and cooperation
- Promote advancement of electrical industry
- Save TIME and MONEY by having the latest information

PLEASE PRINT

Name - Last   First   M.I.

Chapter, where you live or work, if known

Title

(Division, where appropriate)

Employer

| For Office Use Only | Section No. | Chapter No. | Division No. |

Address of Applicant

If previous member, give last membership number

City   State or Province   Zip or Postal Code

and last year of membership

19____

Student applicants give school attending **

Graduation Date   /   /

Endorsed by

Endorser's Membership Number

Applicant's Signature

(Area Code) Telephone Number
___ - ___ - ___

☐ Check ☐ Money Order ☐ Visa
☐ MasterCard ☐ Amex ☐ Discover

Name on Card

Charge Card Number

Exp. Date

☐ Inspector $40   ☐ Chief Electrical Inspector * $40   ☐ Associate $40   ☐ Student ** $30

Inspector Members and Chief Electrical Inspector Members MUST sign below:

☐ I, _____ , meet the qualifications for inspector member type as described below.

Inspector members must regularly make electrical inspections for preventing injury to persons or damage to property on behalf of a governmental agency, insurance agency, rating bureau, recognized testing laboratory or electric light and power company.

*Chief electrical inspector members must regularly provide technical supervision of electrical inspectors.

**Student member must be currently enrolled in an approved college, university, vocational technical school or trade school specializing in electrical training or approved electrical apprenticeship school.

Mail to: IAEI, 901 Waterfall Way, Suite 602, Richardson, TX 75080-7702   For information call: (800) 786-4234 (8-5 CST)

IAEI

901 Waterfall Way, Suite 602 • Richardson, TX 75080-7702
Phone (214) 235-1455 • Fax (214) 235-6858

Membership Application

Section, National, International,
Sustaining and Inspection Agency

Organization

Name _____

Address _____

_____

City _____ State _____ Zip _____

Phone _____ Fax _____

National, International and Sustaining

applicants may designate one representative in each IAEI Section. (Give name and address)

Eastern _____

_____

_____

Western _____

_____

_____

Southern _____

_____

_____

Southwestern _____

_____

_____

Northwestern _____

_____

_____

Canadian _____

_____

_____

Section Member Applicants

may designate one representative in their IAEI Section.

Name _____

Section _____

Membership Category & Dues

Check the one box below that best describes the membership category of your application. Please see the membership category descriptions on back of this page.

☐ Section Member .... $60.
☐ National Member .... $220.
☐ International Member .... $300.

Sustaining Member

☐   Bronze .... $500.
☐   Silver .... $1000.
☐   Gold .... $2500.
☐   Platinum .... $5000.+

☐ Electrical Inspection Agency .... $80.

Electrical Inspection Agency Applicants

may designate one Inspector Member as their representative.

Name _____

(Representative must qualify as Inspector member in Articles of Association.)

Applicant

Signature _____

Title _____

Endorser (New Apps) _____

Member No. _____ Date _____

Chapter _____

This is a ☐ NEW ☐ UPDATE Application

IAEI 1996 Membership Classifications

Section Member. An organization or firm that operates within the territory of a single Section and is interested in promoting the objectives of the IAEI. A Section member may have one (1) representative in the Section where their operations are conducted. Such representative shall be entitled to the privileges of an Associate member unless qualified as an Inspector member.

National Member. An organization or firm that confines its operations to a single country and is interested in promoting the objectives of the IAEI. A National member may have one (1) representative in each Section within the boundaries of the country where operations are conducted. Such representative shall be entitled to the privileges of an Associate member unless qualified as an Inspector member.

International Member. An organization or firm that operates on an international basis and is interested in promoting the objectives of the IAEI. A member in this group may have one (1) representative for each Section. Such representatives shall be entitled to the privileges of an Associate member unless qualified as an Inspector member.

Inspection Agency Member. A municipal, county, state, provincial or federal government or private inspection agency which makes inspections of electrical materials, devices, appliances or installations, for the purpose of preventing injury to persons or damage to property, which is interested in participating in attaining the objectives of IAEI. Such membership shall provide one Inspector membership for the electrical inspector principally engaged in electrical inspection for the inspection agency.

Sustaining Member. An organization, firm, association or group which is particularly interested in supporting the objectives of the IAEI by paying dues not less than that shown in this section. Such Sustaining member may have one (1) representative for each Section. Such representatives shall be entitled to the privileges of an Associate member unless qualified as an Inspector member. Sustaining memberships are further established as bronze, silver, gold and platinum types.

MA1996S

3-11-97